THE SCOPE OF GEOGRAPHY

RHOADS MURPHEY

THE SCOPE OF GEOGRAPHY

Second Edition

RAND McNALLY COLLEGE PUBLISHING COMPANY • Chicago

Rand McNally Geography Series
Edward B. Espenshade, Jr., ADVISORY EDITOR

SECOND EDITION

Printed in U.S.A.
Library of Congress Catalog Card Number 73-9086

for
Eleanor

Table of Contents

Figures

Maps

Table

Introduction

THE STUDY OF GEOGRAPHY is a field whose boundaries may seem difficult to define. To many people it may appear that geographers have no coherent idea of what their own discipline encompasses and that they pursue their work by trespassing on other disciplines. It is true that the scope of geography is extremely broad and that the nature of the discipline leads it into investigation of a wide range of matters, most of which are to varying degrees also the concern of other disciplines. But there is a clear set of concepts which occupy the core of geography and which direct and shape all of its investigations. These are concepts which are not part of the structure of any of the other social sciences. However much the geographer may appear to be treading on other people's toes, he is in fact pursuing questions which, generally, no one else is asking. This book attempts to examine these questions, and to present a coherent picture of what the discipline of geography is all about—its nature and its scope.

A brief may nevertheless be argued for breadth of interest, in addition to the essential focus of a discipline. The study of human society or of the physical world belongs, after all, to no discipline, and to all. Wide horizons are personally rewarding, but such perspective is also desperately needed in a shrinking and conflicted world. Understanding of other places and cultures[1] is vital in the modern age, more so than any scientific or other knowledge. Geographers are fortunate in that their practice may involve examination of a very wide sample of the human experience as a whole. History, anthropology, and sociology also have a broad concern with the human experience, but none of them has so extensive a concern as that of geography. It is the perfect profession for the curious.

Curiosity alone, however rewarding the path it dictates, is not enough; it must be focused and organized if it is to yield anything more than personal satisfaction. Information by itself does not produce knowledge; conceptual analysis is necessary. That is what "discipline"

[1]The word "culture" in its general sense, as used, for example, by anthropologists and throughout this book, means the total way of life of a people, including their system of values, system of social, political, and economic organization, means of livelihood, land use, language, material structures, clothing, diet, techniques, and art forms.

means, in academic terms: the investigation of phenomena in accordance with a previously developed set of concepts and techniques which together form a coherent whole and can give useful meaning to the study of the materials it chooses. A discipline must ask particular questions and seek particular understandings. However wide its search, it needs to be able to discern patterns, relationships, and the manifestation in the real world of those theoretical constructs which give it significance as a field of study.

A discipline whose nature, like geography's, leads one into the broadest possible study of human society, in its terrestrial frame, brings special rewards. The geographical, or spatial, dimension is involved in the great majority of man's affairs—past, present, and future. Any study with such a scope is intellectually demanding and requires a formidable amount of knowledge as well. If geography is the profession for the curious, it is also one best practiced by a truly educated man—and that is the reward: a combination of discipline and the freedom implicit in a study which takes so much of the human experience as its field. The chapters which follow aim to show how this is accomplished.

1
Geography and Social Science

"GEOGRAPHY" COMES FROM a Greek word meaning literally "description of the earth." But modern geography is concerned with man as well as with the earth and with relationships and analysis as well as with description. The geographer analyzes the physical world and examines relations between places in order to throw light on the pattern and nature of human society. He investigates the interrelationship which exists between man and his physical environment. He examines regional differences, attempts to account for them, picks out regional patterns, and tries to draw regional lines and to discern regional relationships. The geographer concentrates this study of the earth and its spatial framework, or the pattern of distribution of things on the earth's surface, toward a better understanding of the human world. He sets man in the framework of the earth which he inhabits.

The broad scope which is necessary in dealing with the analysis of spatial patterns over the world and of man's varying use of the earth involves the geographer with a great number and variety of facts. By themselves, as individual pieces of information or simple descriptions of places and features, facts are less useful than if they can be related and where possible used to reach conclusions. Information does not equal knowledge. It is only when facts can be put together to form a pattern or to support a general statement of relationship, or when conclusions can be drawn, that progress is made toward understanding. The assembling of separate facts into a pattern involves the process of generalization. Generalization, or abstraction, is seldom easy, because it tries to go beyond the individual uniqueness of facts or single phenomena to discern their broader common features, relations, or causes. Language itself is an abstraction, and practically all words in any language represent generalizations which ignore individual variations, for example, "table" or "man." Few generalizations are safe from attack unless they are so general as to be useless, but understanding suffers unless the attempt is made to construct a larger frame of common features into which separate facts may be fitted. To generalize about broad patterns or about cause and effect relationships in human society is especially difficult because the problem is highly complex. Generalization can be misleading, as it glosses over exceptions and contradictions. But just as the individual learns by generalizing from his accumulated experience, so the scholar must learn to abstract patterns and to reach conclusions about the world from the variety of accumulated facts about the world. The dangers of generalization are clear, and every student, even at a beginning level, is under the obligation to make a careful and reasoned use of facts. But he must make use of the facts and not gather them as if they were ends in themselves. Facts never "speak for themselves"; they have meaning only when they can be related or grouped into some sort of system.

This book attempts to guide the student toward the use of certain facts by means of the body of ideas which are employed in the field of geography. Since any one book is limited, and since this one attempts to introduce the student to a large field of knowledge, it necessarily deals in a great number of broad generalizations. Many of these can be challenged by pointing out exceptions or contradictions, although it is hoped that all of them have enough validity to stand in general terms. In a scholarly treatise, exceptions and contradictions must be examined with care and all the relevant evidence and possible conclusions presented. It is usually not possible to write a general introduction like this book as if it were a scholarly treatise, nor is it necessarily desirable. Once the dangers and limitations of abstraction are made clear, the student must be encouraged to develop a conceptual framework and a pattern of ideas with whose help he can analyze, relate, and interpret certain facts. To begin with, he can usually do this most usefully in broad terms. Clarity and understanding are usually enhanced by attempting to see the pattern of the forest as well as the individual trees.

GEOGRAPHIC ANALYSIS

It is probably true that each of the intellectual or academic disciplines is primarily an attitude of mind. Each of the social sciences represents one approach among many to the common objective of knowledge about mankind, and the distinctiveness of these various approaches is more a matter of what each is looking for than of a set of mutually exclusive facts, tools, or methods. The geographer looks for particular factors and particular relationships to which he is attracted by a particular attitude. The same material may also be examined by the historian, for example, or the economist, just as the geographer may make use of historical or economic data. Broadly, the geographer is looking for spatial form and spatial relations or, in other words for patterns of distribution and interaction. The other disciplines are less concerned with these patterns and are therefore less likely to discover them, although they may find the geographer's conclusions useful toward their own purposes.

There is a necessary division of labor involved among allied disciplines, but it should not be assumed that the need for collaboration is thereby avoided. Circumstances often make it necessary for their coverage to overlap, but this does not destroy their respective identities or responsibilities. An effective arrangement depends on broad collaboration as well as on distinct functions. Distinct areas of analysis explored by geography may be summarized under three main headings as follows:

1. The distribution and relationships of mankind over the earth and the spatial aspects of human settlement and use of the earth. (Chapters 2 and 3)
2. The interrelationship between human society and the physical environment as part of the study of areal differences. (Chapters 4 and 5)
3. The regional framework and the analysis of specific regions. (Chapter 6)

The reader may complain that "space" is a very broad or even ambiguous word. Indeed, all three of these headings might be lumped together as spatial study. The division of the earth into regions and the analysis of specific regions are both concerned with spatial form and distribution. Examination of differences in human society from place to place as they may be related to differences in the physical environment, and of differing landscapes as they reflect a changing human agency, are also concerned with earth space. Environmental factors such as mountains, rivers, or deserts affect relations between one place and another and thus also affect the nature and distribution of man's activities, which may in turn transform or severely damage the physical environment. Geography is basically concerned with the patterns of man's distribution over and use of the earth, why he uses it differently in different places or at different times, and how this is related to the whole of his cultures and economies. These are broad complex questions; many approaches in various fields may be used toward answering them. The geographer must depend not only on his own analysis but also on the approaches and findings of others. But the phenomena which he chooses to investigate, those in which matters of spatial distribution or the physical environment and landscape or the region are involved, lend themselves to geographic analysis, and in this sphere he has the principal responsibility for assessing the manifold lines of evidence and suggesting conclusions.

The geographer, like other social scientists, believes in cause and effect. The pattern, organization, and nature of human society are not haphazard. The reasons are seldom simple, and some of them can be discovered only in part or not at all. But this does not invalidate the attempt to discover them nor make it a waste of time when the attempt fails to produce definitive or measurable results. Human behavior is lawful. Unless we assume this, there seems little purpose in trying to understand it. If we do not ask questions, we do not learn. And if we believe in the importance of causality, we must also believe in the applicability of principles or concepts. *Laws* is too extreme a term and may never be fully appropriate, at least not in the sense that the chemist can use Boyle's law or the biologist can use the Mendelian laws of heredity. Human society is less regular than the physical world and it cannot yield the same kinds of results to the student. The theoretically lawful behavior of society is the result of almost infinite variables as causes. It cannot be understood or predicted as well as the behavior of individual people, let alone the behavior of physical matter. Our aim is to introduce a degree of order, to establish relationships, patterns, associations of factors. This is seldom easy. Because phenomena are associated or occur together, in space or in time, does not necessarily mean that they are causally related. It may mean this, or it may not. Such an association is more likely to be part of a much more complex relationship among a great variety of *interacting* factors, each influenced by the presence of the others. No one of them alone is both necessary and sufficient to produce the observed result, but taken together they may form a causal association or a consistent pattern which may recur under similar conditions elsewhere, although it is extremely rare for it to occur in exactly the same way twice. So many different and interacting factors are involved in most aspects of human society that the necessary set of circumstances is almost never duplicated. But although each area, each historical period, and each human phenomenon is different, causation is universal and certain common features can be found. Without this kind of analysis, there is merely an accumulation of data, which by itself may often be meaningless. The student of society must be concerned with these and other complex structures of causality and must attempt to distinguish simple coincidence or association from cause and effect.

SCIENCE AND SOCIAL SCIENCE

It is seldom possible in the social sciences to achieve the precise results frequently characteristic of the physical sciences. This lack of precision is not so much because the study of society, in its modern form, is comparatively recently developed as because much of its material does not easily lend itself either to accurate measurement or to controlled conditions of observation and experiment. Human society is a vastly complex affair, and it is always changing, always in motion. Almost nothing about it is constant, so that for each point in time a different combination of factors, known or unknown, and a different pattern of effects confront the student. He must distinguish qualitative changes as well as more easily measured quantitative changes, whether he is attempting to study a particular process or simply an aspect of society. Even without regard to change, material which he may arbitrarily attempt to isolate for examination may be quite unrepresentative, however carefully it is chosen, since he is dealing not with inert matter, which usually behaves regularly, nor even with biological matter in which one red blood cell is like ten thousand others, but with something far less orderly or fixed, such as a city or a system of land use. Each specimen is likely to be unique. He may speak of examining it, but he cannot take it into a laboratory; he cannot manipulate it but can only observe it. And however small the problem may be, it is impossible, without distortion, to isolate any aspect of human society from its manifold interconnections with other aspects.

Some things about human society, and some aspects of some things, may be more or less precisely measurable under favorable circumstances, for example, population totals for definite areas, or value of recorded foreign trade. But many things and many aspects are not, or are measurable only in small part or by rough approximation, for example, the bases of political power, or regional differences in culture (see note 1 to the Introduction), or population totals and trade flows in an illiterate subsistence society. It is clear that the student of society must depend upon more than quantitative methods, although quantitative analysis can be very helpful to him where it is appropriate. Geography, especially, by its very nature incorporates an enormous field of study, encompassing, like history, most of human experience. The geographer must depend on a variety of methods in his analysis and cannot limit his investigations either to strictly quantitative measurements or to purely subjective or intuitive judgments.

In many ways it would be preferable to avoid the word "science" in connection with the study of society and to divide all learning into *humanities* and *sciences*, the first concerned with the nature and works of man, the second with the nature of the physical world. Most systematic inquiry into the nature and works of man, however, is best undertaken by means of the logical method, or, as it is sometimes rather ambiguously called, the "scientific method." Logic is far from being the exclusive property of the physical sciences, and indeed it has been most highly developed in one of the humanities, philosophy. Rational analysis of any kind depends on logical inquiry. The physical sciences differ from the social sciences or the humanities mainly in the nature of their material. Experimentation is a major basis of the logical method since only thus can alternative hypotheses be tested until the most suitable one is established. But experimentation in the laboratory and test-tube sense is impossible in most analysis of society. The student of society must be content by and large to apply alternative explanations and conceptual systems to social situations as they exist and to see whether he has really explained all the facts. With the interlocking complexity, awkwardness, and volatility of his material, with his inability to isolate or manipulate it, and with the handicap that he is far from having all or perhaps even most of the facts about mankind, it is understandable that he cannot compete with the accuracy and finality of the physical scientist.

Actually, the physical scientists are themselves finding that the physical world is less determinate and conforms less to a regular fixed order than has previously been assumed. Many of the long-accepted "laws" of physics, for example, have been disproved, and many of the seeming regularities in nature have been shown to be misleading. In this sense, there are no "natural" laws, but only man-perceived regularities, patterns, or associations. Students of society are, however, still more dependent than physical scientists on qualitative rather than quantitative methods, and they must often make value judgments. This does not make the study of society less interesting or less important, but it does make it different from precise, determinate science. Literal minds which like to think in black-and-white terms or in terms of fixed "laws" are not well suited to theoretical science of any kind, and certainly not to the study of society. The student of society believes in the scientific or logical method, but in terms of his job this cannot be more precisely defined than *the careful use of facts, reason, and systematic thought, aimed at the discovery of patterns or associations.*

PREDICTION AND CAUSALITY

Prediction is a somewhat controversial matter among students of society. It would be foolish to maintain that prediction is either entirely impossible or undesirable. But it is plain enough, in terms of the foregoing discussion, that most students of society are not in a position to read the future with precision except in limited cases. Prediction is an important function, and may in certain cases be extremely valuable. But the social sciences, certainly geography, do not stand or fall on their ability to predict accurately, nor should that be their chief objective for the future. Their other work is not invalid because it does not deal with or produce laws of complete predictive value; it is more appropriate for most analysis of society to speak in terms of probabilities or recurrent patterns. But when the social scientist is so far from laboratory conditions in dealing with the present or the past, it is not to be expected that he can do any better, if as well, for the future, when many of the assumptions he makes and the factors underlying the patterns he observes in the present or the past may be radically changed. Human society is never constant, and although it is part of man's function to pick out patterns of its inconstancy, prediction is only an incidental part of that function. He is concerned with patterns and associations in human society as he is able to observe it; if this analysis helps to suggest patterns of future development, so much the better. Perhaps in time, as he learns more, some of these suggestions may begin to have greater predictive value, and no one would be happier about that than the student of society. But prediction will not then become his main function. He will still have more than enough important and useful work to do toward understanding human society as it is and as it has developed, which is the principal approach of this book.

It should not need arguing that any educated man should understand as much as possible of the world society in which he lives. An understanding of man's distribution over the earth, his fragile interrelationship with his environment, the different uses to which he has put different parts of it, the cultures and economies he has created, and the spatial interrelations which exist between and have influenced these patterns—or in a word, geography—has a fundamental place in the equipment of an educated man.

In the search for consistent patterns and associations, the social scientist is seldom in a position to say, as the physical scientist often is, "If *a* is present, then *b* will follow." Mankind is not that simple, nor does he know enough about human society or about the interlocking factors influencing it. He is more likely to be able to say, "If *a*, *b*, and *c* are present, then *d*, *e*, or *f* may also be associated with them, but not g, *h, i, j,* or *k*." He can usually eliminate irrelevant factors and can usually determine the set of factors or conditions among which to seek a solution, but not often much more. He must also assume, as a result of accumulated experience, that wholes or aspects of societies or economies can rarely be explained adequately in terms of a single factor. The very complexity of human society and its interrelations, which makes simple explanations inadmissible, also makes a continual temptation to cut through the confusion, have done with the if's and but's, and devise a simple direct idea or set of ideas which can consistently explain everything. There are many famous examples of this kind of fallacious analysis. Marxism, especially as it has been used by people less principled or objective than Marx, is an outstanding one. There are many theories of history open to the charge of single-factor oversimplification. In the field of geography, the single-factor heresy is represented by environmental determinism, which maintains that everything, or nearly everything, about mankind is directly traceable to some aspect of the physical environment.

The damage which these misguided enthusiasts do is very great but it is often obscured by the insights which they simultaneously create. They successfully dramatize an idea, usually a most important one, or develop a theory which throws new light on a major problem. The more useful and interpretive they make the idea the harder it is to make clear that far from simplifying the problem they are often confusing it further by attempting to disguise its complexity. This makes an adequate explanation more difficult, especially when it must compete with the dramatic appeal of a "simple" answer by a weary recitation of complex problems, complex intercausation, and complex qualified solutions. The proper student of society can proceed only on the assumption that all single-factor explanations are to be distrusted if not rejected, and that he must always preserve (1) an open mind about what may be important or relevant and (2) a stock of multiple concepts, or a readiness to formulate new ones, for every situation which he investigates. He must try not to be burdened with preconceived theories into which he attempts to squeeze ill-fitting data, but to use data to construct and to prove or disprove theories which he is always prepared to modify or discard. Generalizations are not developed in order to construct a theoretical system as an end in itself. The goal is the fuller understanding of man and his works.

Progress toward such understanding is not well served by pursuing any set theory to the exclusion of others or by concentrating on the use of any one method. This error appeared in the discipline of geography at one stage in its development in the form of environmental determinism. In the present generation it is important to resist the assumption that all of human society can be explained, or predicted, by quantitative analysis, or that all of geography can be made an exact "science." But it is equally important for the geographer not to ignore the question of causality. He must not merely gather and arrange data or write description; he must continually ask Why? In this endeavor, multiple methods and multiple ideas are relevant and necessary for fruitful results.

This discussion of causality in the study of society is not intended as a digression. The chapter began by stating the major areas of geographic analysis and pointed out that in all of them the geographer is looking for associations and asking questions. It has been necessary to summarize some of the conditions under which he is obliged to operate in trying to pick out associations and to answer questions. We must now turn our attention to the questions themselves.

Selected Samples for Further Reading

Benedict, R. *Patterns of Culture.* New York, 1934. A little classic, by a late leading anthropologist, which discusses the nature and diversity of culture, its importance as affecting human actions, and its relevance among three widely differing peoples, as compared with the much more complex European-North American cultural milieu.

Blalock, H. M., ed. *Causal Models in the Social Sciences.* Chicago, 1971. A collection of essays on causality, prediction, and quantitative techniques as applied in a variety of disciplines.

Brock, J. O. M. *Compass of Geography.* Columbus, Ohio, 1966. A well-written succinct presentation of the essence of the geographic approach and its historical development.

Brookfield, H. C. "Questions on the Human Frontiers of Geography," *Economic Geography,* XL (1964), 283–303. A critical review of recent literature, stressing the importance of the social dimensions in geographic inquiry.

Brown, R. *Explanation in Social Science.* Chicago, 1963. Considers the nature of social science and the degree to which a truly scientific explanation of society is possible.

Bunge, W. *Theoretical Geography.* Lund, 1962. The nature and place of theory in geographic inquiry.

Chisholm, M. *Geography and Economics.* New York, 1966. How geographers and economists have, jointly and separately, analyzed the causes of economic phenomena—a brief and clearly written survey of a close interrelationship.

Cohen, M. *Reason and Nature: An Essay on the Meaning of the Scientific Method.* New York, 1931. Old, but still one of the best statements, among many, of the problem discussed in this chapter.

Dickinson, R. E., and Howarth, O. *The Making of Geography.* Oxford, 1933. The historical development of the discipline of geography from classical times to Humboldt.

Freeman, T. W. *A Hundred Years of Geography.* Chicago, 1962. The development of the discipline since the mid-nineteenth century.

Geyl, P., et al. *The Pattern of the Past: Can We Determine It?* Boston, 1949. A stimulating collection of essays on an old problem, with many applications to the field of geography.

Haggett, P. *Geography: A Modern Synthesis.* New York, 1972. A sprightly new introductory text which pursues and gives imaginative illustrations of the discipline's three major concerns with space, the environment, and the region.

Harvey, D. *Explanation in Geography.* London, 1969. A review of some of the philosophical and functional problems of the discipline.

James, P., et al. *American Geography: Inventory and Prospect.* 2nd ed. Syracuse, 1964. A collection of summaries of the state of progress in a variety of subfields within geography, with some more general considerations.

Kuhn, T. S. *The Structure of Scientific Revolutions*. Chicago, 1962. A general introduction to the dynamic evolution of physical science.

Mikesell, M. W., and Wagner, P. L., eds. *Readings in Cultural Geography*. Chicago, 1962. A selection of articles in the broad field labeled "cultural geography," with useful and stimulating, if not wholly definitive, introduction and commentary by the editors.

Redfield, R. "The Art of Social Science," *American Journal of Sociology*, LIV (1948), No. 3. The importance of perceptive understanding and significant generalization depends on something more than formal method of quantitative analysis. This article discusses imagination and intuitive judgments and the connection with the humanities.

Schaefer, F. K. "Exceptionalism in Geography: A Methodological Examination," *Annals of the Association of American Geographers*, XLIII (1953), 226–49. One view of the comparison between geography and "science."

Schmidt, P. F. "Some Merits and Misinterpretations of Scientific Method," *The Scientific Monthly*, LXXXII (1956), 20–24. The title is self-explanatory.

Soja, E. W. *The Political Organization of Space*. Washington, 1971. A general approach to human territoriality.

Spate, O. H. K. "Quantity and Quality in Geography," *Annals of the Association of American Geographers*, L (1960), 377–394. Telling criticism of some of the shortcomings of quantification in geography, and a persuasive plea for a discriminating balance.

Spencer, J. E., and Thomas, W. L. *Cultural Geography*. New York, 1969. A recent text, the first to attempt a comprehensive survey of this highly varied subfield.

Taaffe, E. J. *Geography*. Englewood Cliffs, N. J., 1969. A brief account of the current state of the discipline, stressing a variety of applied aspects.

VonWright, G. H. *Explanation and Understanding*. Ithaca, 1971. A useful philosophical and methodological approach.

2
Spatial Form and Spatial Interaction

THE FIRST FUNCTION of geography is listed in the preceding chapter as a study of "the distribution and relationships of mankind over the earth and the spatial aspects of human settlement and use of the earth." The geographer wants to know how and why man is distributed as he is: what accounts for the great clusters of population and for the empty areas, for the cleared and uncleared land, the cities, the villages, and the roads, what spatial patterns can be observed, and what interactions there may be between them.

If the roughly 3.7 billion people of the world in 1972 were distributed evenly over the earth's land surface, there would be an average population density of about seventy people per square mile. If they were gathered together as closely as an audience in a theater, three hundred square miles would hold them all. As it is, about half of the earth's land area supports fewer than two people per square mile, some densely populated rural areas support as many as three thousand per square mile, and population density in the centers of large cities may reach or exceed a hundred thousand per square mile. Settlements of various kinds are scattered widely over the earth, but only a few have become big cities. In some rural areas of the world, farmers live in villages (nucleated settlements), and in others they live in isolated farmhouses (dispersed settlements). The tea shrub, like many other commercially useful plants, will grow in many different places, but only a few areas actually grow tea on a large-scale commercial basis. Mineral ores and fuels occur in scattered locations over much of the earth, but heavy metal manufacturing has developed on a large scale in only a few places. There are approximately three thousand different languages in the world; some are spoken only by a few people, some by hundreds of millions, some are limited to a single large or small area, and others have spread over widely separated areas. What are the implications, causes, and interrelations of these varied spatial patterns?

Geography is concerned wherever the fact or the concept of *place* or the question *where* is involved. Places and spatial patterns, as in the examples cited above, are not interpreted as static phenomena or simply as facts, but as concerned with different sets of relationships. It is only of incidental importance where Timbuktu is (or China, or the wet tropics, or the area where Swahili is spoken), except as a knowledge of the location helps to understand the relation between Timbuktu and the rest of the world and the effects which these relations have had on Timbuktu and vice versa. No place and no pattern of distribution in the world is exactly like any other place or pattern. There are also marked similarities among groups of places and spatial patterns. In each case, the nature of relations with other places may be fully as important as local conditions in accounting for the differences or the similarities. The geographer attempts to analyze human society in its spatial frame. He examines not only where things are and what forms or groups may be discerned but why they are there and how their location and their nature and development may be related to the influences exerted by other places. This is the meaning of the term *spatial interaction.* Spatial form is significant only when it can be related to spatial interaction.

The development of theory which can help to explain real spatial patterns can often be aided by imagining an unreal featureless earth as smooth as a billiard ball on which people and resources are evenly distributed. By adding the varied elements of complex reality one at a time to this simplified image, the consequences of each element may be deduced, and some of the aspects of experimental science simulated. Location theory, in particular, makes use of this technique as in the development of the theory of the *central place* discussed below. The understanding of complex reality requires the kind of generalization which theory makes possible and out of which theory is constructed. But generalization, and the oversimplification on which it necessarily rests, must continually be tested against specific pieces of reality.

SPATIAL RELATIONS

Timbuktu

Timbuktu is a real city, despite its unreal overtones. Why is it where it is, and why has the settlement on this site grown big? (Consult an atlas map of North Africa.) Location on the Niger River is clearly relevant, for Timbuktu lies near the margin of the Sahara Desert, and, in the absence of adequate rainfall or lakes and of ground water for wells, settlement in this area must be limited to sites where water is available. *Site* refers to the actual ground on which a city or other phenomenon rests. It is often paired or contrasted with *situation,* which refers to spatial relations with other places, principally in terms of accessibility rather than of simple distance. Situation is thus both a relative and a changeable matter. Timbuktu's situation—its relations with other places—has been much more important than the local conditions of its site in accounting for the city's growth and decline. Northward is the desert. Southward the land becomes more humid and therefore more densely populated, with a type of productive economy different from that of the desert. Settlements which lie in the zone of transition between one set of physical or economic conditions and another set may be convenient centers for the exchange of the different sorts of goods which each area produces. Timbuktu has prospered as a trade center in part for this reason. The city also has the advantage of easy access to the moister and more productive area to the south by means of the Niger River, which flows southward to the coast. Towns and cities are specialized concentrated settlements which have grown primarily because they have access to wider areas around them and thus can perform on behalf of these wider areas economic, political, or social functions which are most conveniently carried on in a central place, such as government, marketing, manufacturing, or the servicing of trade routes. The differences in the physical nature of the areas to the north and south served by Timbuktu promoted a *complementary* ("filling mutual lacks") trade relationship between them, since they produced and needed to exchange different kinds of things. Regional differences do not always result in exchange. The amount of exchange depends on the effective demand in one region for the commodities produced for sale in another region and on the cost of transport between the two places.

Timbuktu is a convenient center through which the complementary exchange between the different areas around it can take place. This exchange is due not only to its median location, but because transport carriers are likely to change there, from land to river, from desert transport to humid-area transport. Timbuktu has in fact been called "the place where camel meets canoe." Places where carriers change, or *break-in-bulk points*, will have economic opportunities for the support of their populations. This will apply not only to loading and unloading, storage, or servicing of transport carriers but also to the financing and insuring of trade and to the processing or manufacturing of goods passing through. The goods must be unloaded at the break-in-bulk point; transport and loading-unloading costs for those goods are therefore minimized at that point, and a variety of routes are likely to focus there, funneling goods through the city. These are important matters for any manufacturing or service industry dependent on the assembly of raw materials or the distribution of finished goods. The trade through Timbuktu has never been very great, by modern standards, because the tributary areas which the city serves are sparsely populated and unproductive (the desert), or largely organized economically on a subsistence rather than a commercial exchange basis (the moister areas to the south). Hence Timbuktu is not a very big city, but it is the biggest and almost the only city in this part of northwest Africa.

Timbuktu would be very much smaller, however, if it were only the desert which lay to the north. The desert produced almost nothing for exchange, but beyond it is the moister Mediterranean coast, which did produce something different from the areas south of Timbuktu. This coastal region also exchanged goods with humid central Africa over the most practicable route, which goes through Timbuktu. Some of these goods were redistributed from North Africa by cheap sea transportation to wider markets in Europe and elsewhere, where under different physical and cultural conditions the same goods, such as ivory, gold, spices, or slaves, were not produced. These wider relations were reflected in the size and functions of Timbuktu, hundreds or even thousands of miles away. The goods which moved along the route north of Timbuktu had, however, to be of a certain sort. Here there was no river to lessen transport costs, and it was a very long haul to the Mediterranean coast, through an area largely empty of population and devoid of opportunities for exchange. The carriers (camels) are very expensive, measured by the unit cost of transport, or the cost per ton-mile (for carrying one ton one mile), mainly because their individual capacities are small. Transport costs tend to vary inversely with the capacity of the carrier, which is the principal reason why water transport is so cheap. The length of the route between Timbuktu and the Mediterranean coast increased the expense, as did the lack of trade-generating centers along the way. The latter are important factors in making a transport route profitable. Therefore this route could carry only goods which were very high in value by weight, such as gold, ivory, slaves (who provided their own transport), or salt (which is very high in value in most preindustrial economies).

With the development of mechanized sea transport and the widening of Africa's commercial relations with the rest of the world, the goods which used to move through Timbuktu to wider markets began to move instead through west African seaports. Timbuktu has become a much smaller place and more exclusively dependent on local trade and services. Change, especially technological change, is continually altering spatial relations and spatial interaction patterns. At its height about the fifteenth century, Timbuktu was a commercial metropolis. Because of its size and its excellent access, it was also a famous cultural center for much of the large area of the western African Sudan, and supported a well-known Moslem university. With the deflection of the trade routes which fed it and their interruption by political disorders, Timbuktu's glories faded. By the end of the nineteenth century, the French found it a largely ruined town. Both political and technical change were thus responsible for Timbuktu's decline, as they affected the city's relations with other places or its pattern of spatial interaction.

Other Examples of Spatial Interaction

The British economy and society have been profoundly influenced by Britain's access to and from the rest of the world. Similar physical conditions exist in many other parts of the world, but the particular sort and degree of accessibility which Britain enjoys cannot be duplicated. This suggests that the country's human uniqueness may be traceable to its major locational uniqueness. Other factors which have little or nothing to do with location have also been important, but a study of the locational or spatial factor can provide considerable insight toward understanding Britain. California's agriculture is now specialized in the growing of fruit and vegetables partly because of a favorable local climate, but also because in the course of the last seventy-five years California has developed cheap, easy, and quick access to a large market in the eastern United States, where the climate is different. Without the east, if it were a different and less complementary kind of place, or if it were not for the railroads, California agriculture would be a smaller and very different enterprise. The effects of weak spatial relations may be equally important and obvious in other situations. Isolation was as responsible as environment for the uniquely primitive level of the Australian aborigines and for the low level of technical development characteristic of the North American Indians and of the various groups native to the rainforests of Africa and South America.

Analysis of the relations between places can be used to throw light on every problem which the geographer investigates. It is part of his principal function, the study of space and place. He must attempt to see a region, or the world, as a whole. Comparison is also an essential part of geographic analysis and interpretation. The two great poles of world civilization, Europe and the Orient, cannot be fully understood until they have been compared. Comparison helps to reveal the distinctive qualities of each and may suggest some of the distinctive dynamics which without the comparison may be taken for granted or overlooked because the areas have been examined only in their own respective contexts. One's own region or country can even less well be understood and interpreted until it is compared with some knowledge of other regions or countries and thus seen in some perspective. A knowledge of only one area or society may also be seriously misleading if it is applied to the interpretation or evaluation of another area or society. Use of only a single comparative criterion without knowledge of the wide differences elsewhere (and without realizing that there is no single cultural norm) perverts the purposes and benefits of comparison, and it is regrettably common among Americans, whose knowledge of most of the rest of the world is dangerously limited. The comparative method is of course no special property of geography. Anthropologists and sociologists in particular use it as the basis of much of their work, and it is as important in history as in geography. The geographer merely applies the comparative method to the problems of spatial form, spatial relations, and regional differences.

GEOGRAPHY AND HISTORY

The classifying, describing, and interpreting of spatial patterns which geography attempts, including the division of the earth into regions, is the principal basis of the frequently made analogy with history, which classifies, describes, and interprets patterns in time. Every investigation of human society is more useful if it can place its particular inquiry within the double frame of historical time and geographical space. Neither historical nor geographical analysis is complete unless it critically applies both dimensions. Historical changes alter spatial patterns and spatial relations. No landscape and no pattern of land use is static; it is continually changing, and it shows varied and cumulative effects of historical developments in the recent and remote past. A geographic analysis which ignores such factors, or which attempts to treat a landscape or a spatial pattern as if it were static or free from change in the future is a travesty of reality. The examination of such changes, their rates, directions, causes, effects, interrelations, and whether or not they are cyclical or recurrent, is as much the geographer's job as the historian's. Correspondingly, spatial and environmental factors affect historical changes. Historical analysis which ignores such factors gives an incomplete, distorted, or at worst lifeless and meaningless picture, since it neglects the stage on which history takes place. "Geographie without Historie hath life and motion, but at randome and unstable. Historie without Geographie, like a dead carkasse, hath neither life nor motion at all."[1]

A basic function of both disciplines, however, is also the collection and presentation of information—the historical record and the description of regions. Both historian and geographer would deny that their function ends with classification and description. It is unfortunate in one sense that both fields have inherited this function, since both tend as a result to bog down in the time-consuming process of completing the record. History is often identified by long-suffering students with dates and names, geography with capitals, countries, and principal products. So much of both sorts of information is essential for fruitful study of almost any aspect of society that this elementary step is sometimes never left behind. But in both history and geography the collection of data is a means to an end, the end being not the acquisition of information but the development of insight and understanding. The historian attempts to relate and explain phenomena in time, the geographer to relate and explain phenomena in space. Both disciplines are concerned with the "why" as well as with the "where" and the "when."

Geography and history are both *holistic* approaches, in that they are both concerned with the *whole* of human society, in time or in space, or in a given area or period, rather than with individual parts of it. This approach makes it difficult for geographers and historians to be as precisely analytical, to concentrate on processes, or to deal in definite principles as other social sciences like economics or political science, which are concerned with only a part, or with one aspect or process of society. One does not often speak of "principles of history." Geography's somewhat fuller use of particular analytical concepts arises mainly from its concern with spatial interaction, environmental-societal interrelations, and the regional framework. But it is much more like history in its holistic approach than like any of the other social sciences. All other branches of learning necessarily consider and analyze the time dimension and the spatial dimension; they are no special property of history and geography. History and geography also depend importantly on concepts and discoveries of other disciplines, but only history and geography attempt to integrate all knowledge relevant to historical time and geographical space.

---

[1]Peter Heylyn, *Microcosmus, or a Little Description of the World* (London, 1621), p. 11. (Note the use of the literal translation of "geography.")

CITIES AND SPATIAL INTERACTION

To return to the case of the city, perhaps the simplest and clearest example of spatial form and spatial interaction, one may take as given the facts of its existence, its population, number of houses and factories, miles of paved streets, and the size of its waterworks. These can be found in a guidebook. But to explain these facts and to explain the spatial form of the city, one turns first to its relation with other places. The city is as big as it is and specialized in its particular way because it provides services for a particular area dependent on it, the city's *hinterland.* The size of this tributary hinterland is determined by the ease of access to the city, plus the competition of other cities which might also serve the same area (see Figures 1 and 2). The total productivity of the hinterland and the size of its market are matters over which the city can exercise little influence, except as it can improve its own access. It is the hinterland which determines the city's size and specializations. What grows up in one urban center is primarily a reflection of access, of spatial factors over a very wide area, and of the interrelations which take place within it.

The size and productivity of the American Middle West and the easy routes which lead from it through the Mohawk-Hudson corridor are in large part responsible for the size and prosperity of New York City. Provincial residents of Manhattan may find it hard to accept the

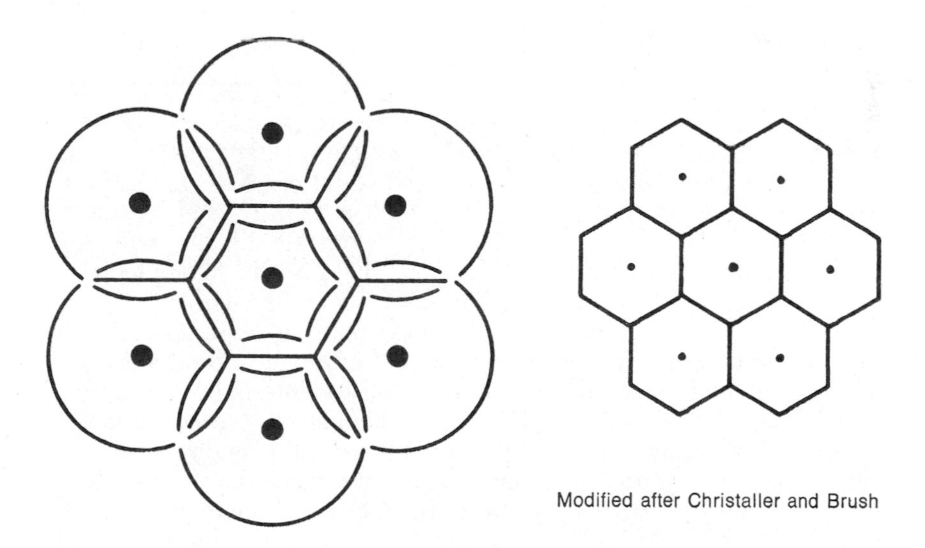

Modified after Christaller and Brush

Figure 1. Hypothetical Spatial Patterns of City Hinterlands.

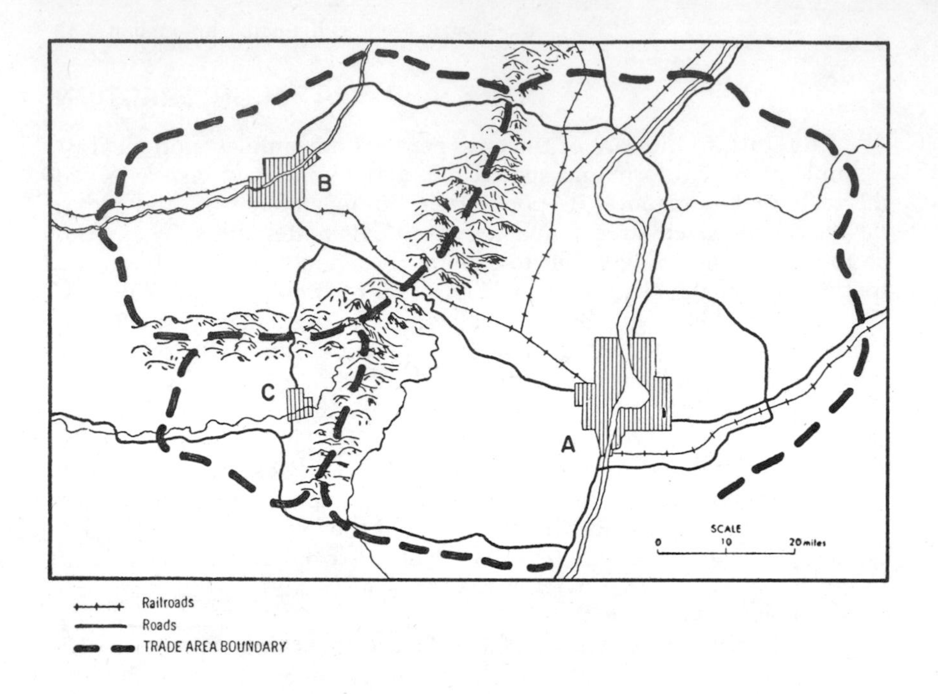

Figure 2. City Hinterlands in a Hypothetical Landscape.

Middle West as a major basis of their existence, but this shortsightedness is characteristic of most city dwellers. In most big cities the butchers, bakers, and candlestick makers for the city's own population employ more people than are employed in the so-called *basic* enterprises, which are producing goods or services to be sold outside. It is these basic enterprises which support the city and make it necessary to have butchers and bakers, cab drivers and street cleaners. It has been estimated that one person employed in a basic enterprise, such as wholesaling, manufacturing, or insurance for areas outside the city, will support approximately two other people employed in the manufacturing and service enterprises supplying him and his family with what they need: food, schooling, department stores, housing, and so on. This roughly one-to-two relationship between employment in production for external sale and employment in production for internal consumption is called the "basic-service ratio." But we cannot all live by taking in each other's washing, even though most city people may do so, and the concept of basic enterprises re-emphasizes the fact that cities exist in order to perform certain specialized functions on behalf of a wider area. A big city like New York will in fact overlap the hinterlands of other cities because it usually has highly specialized goods or services to offer which can be obtained nowhere else, while for less specialized things a smaller and nearer city will be the center. Specialization of any kind, even in a village general store, may reflect particular local advantages or may be simply the result of bigness, but neither of these can be implemented unless the village or city has effective connections with a hinterland tributary to it.

Various attempts have been made to explain the distribution, location, and size of cities and towns, working from the assumption that urban settlements arise to serve a more or less well-defined hinterland. Attempts have also been made to discern consistent spatial patterns of city location and size, on the assumption that there may be a regular order of distribution. From some of these investigations there has been developed the theory of the *central place*, which argues in part that cities may be spaced at more or less regular intervals, each serving a hinterland around it whose limits are fixed primarily by distance. Where distance is the only factor, hinterlands would theoretically assume the shape of a circle whose center is occupied by a city—or since several competing cities actually occupy the landscape, the hinterlands would be tangent on all sides and would thus have the shape of regular polygons, as shown in Figure 1. But simple distance is rarely, if ever, the only factor affecting access, and in reality city hinterlands are grossly irregular in both size and shape, the cities themselves differ widely in size, and the pattern of city distribution is far from regular, as the physical and human landscapes which they serve are not regular or uniform. Mountains, rivers, plains, seas, deserts, and other features of the physical landscape affect access or spatial interaction. Human factors, differing widely from place to place, and especially technological factors as they may influence transportation and production, also affect access. Differing degrees of complementarity between different regions, on both physical and human grounds, are further reflected in differing degrees of spatial interaction. Political boundaries, cultural differences, or linguistic barriers, among many other matters, affect the flow of goods, services, and ideas, and thus also affect the distribution of cities. Among regions, productivity differs, demand levels differ, and each region tends to require a different arrangement and different sizes of cities to serve it. A large semi-arid area with a sparse population and low productivity, or a large area occupied by a dense but mainly subsistence or self-sufficient population, cannot support as many or as large cities as a smaller but more productive and more commercialized area.

The theory of regularly spaced cities and hinterlands is also disrupted in reality by the clearly observable hierarchy among urban places. A large city with good long-range access, New York, for example, may include in its hinterland most of the other cities in the country, to differing degrees. For large capital loans, exclusive dress fashions, or specialized publishing, New York is in many cases the only supplier, and for these things even Chicago is tributary to New York, as most of the Middle West and the larger area served by Sears Roebuck, for example, is tributary to Chicago. Cities of middle rank may similarly include other smaller cities in their hinterlands. At the bottom of the size scale, a rancher in Wyoming, for example, will obtain the bulk of his goods and services from the nearest or most accessible small town in whose immediate hinterland he lives, but he will also be tributary to Cheyenne (the state capital), to New York for cut diamonds or books, and to Chicago for barbed wire and sheep or cattle brokerage. If his wife wants a new hat, she will probably shop for it not in the small town, but the nearest or most accessible small city, whose hinterland for more specialized goods and services overlaps and incorporates that of the small town.

When a city grows big because it serves a productive hinterland, it acquires more and more specialized functions which then further increase the city's size as they enlarge its hinterland. The biggest cities thus tend to have a self-perpetuating advantage. But for any village, town, or city, tributary hinterlands tend to be different for each type of commodity sold or service performed. Medical services will cover one area, food retailing another, wholesaling another, and sale of heavy manufactured goods another. Where manufacturing becomes highly specialized, even a relatively small city may sell its products over a very wide area. For example, Hershey, Pennsylvania, and Hollywood, California, include large parts of the world in their respective hinterlands in this specialized sense, and the same may be said even of the tiny island of Harris off the coast of Scotland, which is world famous for tweed, even though it is poor and sparsely populated. In all of these cases, urban location and urban growth reflect spatial interaction. Relations between places are affected not only by distance but by differing degrees of access, by the complementarity which results from regional differences of all kinds, and by the simple factor of the size and consequent specialization of urban places, which is in turn a reflection of access.

An oversimplified example may help to illustrate the idea of access and the multiplicity of factors which affect patterns of spatial interaction. In Figure 3 a man living at *a* may choose between the cities at *b* and *c* to obtain the goods and services he requires. *B* is closer in physical distance, but a mountain range intervenes. The shortest way there in distance is by road, but the cheapest route for bulk shipments is by

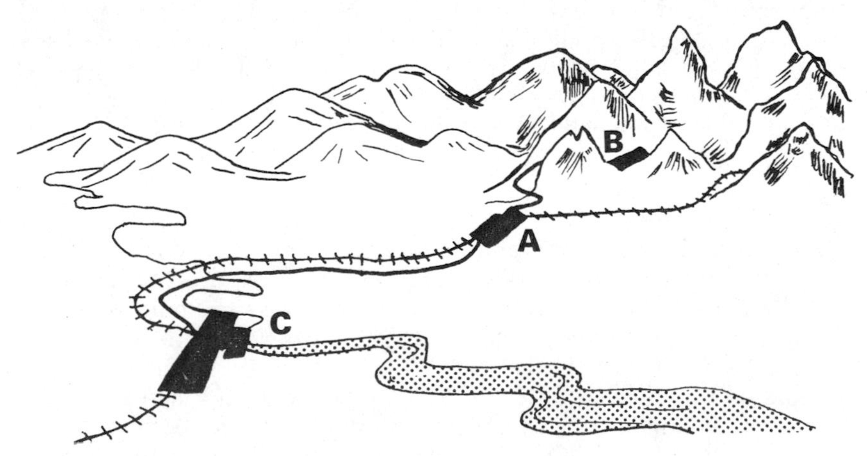

Figure 3. Simple Interaction Diagram.

the railroad over a longer route. If the man has a car and is in a hurry, he will travel to *b* by road, but most of what he orders from *b* will travel via the railroad. There may be other factors, such as the location of intermediate services or shopping places, the enjoyment of scenery, or social and personal benefits which make one route preferable to the other. *C* is a larger city, however, in part because it lies on a productive plain, whereas *b*'s hinterland is restricted in size by mountains and is not as productive per square mile. The man at *a* will therefore find a greater variety of goods and services at *c*, and despite its greater distance will probably obtain there most of what he needs. In reality, the situation would probably be much more complex. Political boundaries, language barriers, or artificial rate structures might intervene which would distort flows between *a* and *c*. Cultural differences between *a*, *b*, and *c*, special local skills, traditional alignments, or differences in law might influence flows in a particular direction. The man at *a* might also quite likely split his spatial relations between *b* and *c*, being tributary to *b* for certain goods and services and to *c* for others. But even in the limited case of the example used, time, distance, costs per mile, intermediate opportunities and rewards along the various routes, the nature of the local and regional environments, and the size and nature of the respective cities, plus other less well-known and more varying factors, all contribute to the pattern of spatial interaction which takes place.

Spatial interaction is by no means confined to cities. Wheat is grown in Dakota and forests are cut over for lumber in Oregon because of the relations between these places and the rest of the world. A century ago wheat growing and lumbering were absent or minor, although the climate, soil, and relief were essentially the same then as now. What has changed is that Dakota and Oregon have been connected by transport networks with highly developed distant markets, especially in the eastern United States and northwest Europe. These markets demand wheat and timber and have the economic and technical means of satisfying their demand from distant sources. They no longer grow most of their wheat and timber locally, although physical conditions there may be as favorable or more favorable for these crops than in Dakota or Oregon. Other activities are now more profitable for Pennsylvania or England as a result of the industrial revolution (see Chapter 13 for a discussion of this term) and of the wider relations between Pennsylvania and England and the rest of the world. Wheat or timber may be obtained from distant sources more cheaply, or local land, labor, and capital may be freed for more productive uses. Wheat yields very well in the climate of northwest Europe or the eastern United States, but wheat is tolerant of climates which are too dry for many crops, and hence it is often pushed out toward the dry margins, leaving the moister climates for more fussy and more profitable crops. But the spatial forms which have grown up at both ends of the line, agricultural, manufacturing, commercial, or others, are in large part the result of spatial interaction. What each area is doing depends on what a variety of other areas are doing and on the low-cost transport which ties them together.

CHANGE AND SPATIAL RELATIONS

One final example of the importance of spatial relations may be cited. For the first two thousand years, more or less, of Western history, the focus of development was the Mediterranean Basin. The Mediterranean is a virtually enclosed sea, relatively free from storms, easy to navigate without losing sight of land for very long, and yet connecting a great variety of diverse regions. Three continents come together there, and the Mediterranean became an easy avenue of exchange very early in Western history when men learned to sail in the Mediterranean's quiet waters. Exchange is necessary for economic specialization and is thus a condition of economic development. Toward the end of the European Middle Ages, however, new techniques of navigation arose. The Age of the Discoveries revealed a world not only larger than the ancients knew but with a definite focus in the Atlantic. The most accessible place was no longer the Mediterranean but the shores of northwest Europe, from which multiple sea routes led to vast and productive new worlds toward both the east and the west. It was this change in spatial relations, rather than the much later industrial revolution, which was mainly responsible for the areal shift of human emphasis which took place in Europe after about 1500. The new pattern of spatial interaction has continued to keep northwest Europe far ahead of the Mediterranean Basin in population and economic development.

Human society is not constant, nor are its geographic relationships. Physical distances and the environment may not alter, but their effects on mankind are changing all the time. The most important reason for this is the changes in human knowledge and technical powers. Man learns to do more and different things with the earth, and thus its resources and its distances affect him differently. Less tangible matters such as human wants, tastes, or cultural and ideological standards also change. Migrations of people further help to spread change and to alter land use and other aspects of cultures or economies. The earth does not mean the same thing to man at different periods, nor does it have the same effect on him, even though some of the major outlines of the relationship may remain relatively constant in a given area. Probably the most dramatic reminder of the dynamic quality of man's relation to the earth and of the impact of changing spatial relations is what has happened in North America since 1492. A land which supported perhaps one million stone-age people in 1492 now supports well over two hundred million of the most technically and economically advanced people in the world. California in 1492 was the home of what appear to have been some of the most primitive of all the American Indians, although its sunshine was just as warm and its oil reserves even larger than they are now.

One may think of any society as occupying a certain stage in the development of its adjustment to its geographic base and to its place in the spatial frame. The progression may be up or down in terms of its development relative to the rest of the world. Civilizations decay as well as grow, and in fact the earliest city civilizations so far as we know occupied sites in or on the margins of the great desert of Eurasia, sites which at that time were highly advantageous but which now support societies marked by relative poverty and lack of economic development. It is not space or the environment which have altered, but man's changing use of them and his ability to make more productive use of other areas whose advantages are now seen to be greater in the light of his new powers and wants.

MAPS AND THE SPATIAL FRAME

The map is the distinctive tool of the geographer, for it is only with a map that spatial relations and spatial form can best be seen and analyzed. Direct observation of any area, even over a long period, can pick out only individual details or small pieces of landscape. Larger patterns can be perceived only when larger areas are seen as a whole, reduced to an observable scale, and many of the almost infinite details of reality eliminated in order to concentrate attention on a few matters in the large. Any map selects only a few patterns to represent, for example, elevations above sea level or distribution of population. An aerial photograph shows most of directly observable concrete reality at a reduced scale in which some large patterns can be seen, but the profusion of detail and the fact that many spatial patterns are not directly observable or do not stand out in an aerial photograph make such a photograph much less clear than a selective map. A map can also represent areal traits such as language distribution, density of population, average annual temperature, incidence of illiteracy, flow of traffic, or distribution of incomes which cannot be seen directly in the large except in small part. Most of the aspects of the earth and of human society can be mapped. Representing them on a map makes it possible to understand them better or from a different point of view, if only because they are placed within the frame of space and can thereby be related to other phenomena which have spatial form. Figure 4 shows a few examples, but see also Figure 25 in Chapter 13. There is a growing variety of mapping techniques to represent different types of spatial form, and there are very few aspects of human society which do not have spatial form. Any atlas makes use of several different mapping techniques, and others are used elsewhere in this book.

Essentially, two kinds of jobs are performed by maps. They show spatial distribution or occurrence (spatial form), and they show relationships. The two are of course interconnected, since the nature and degree of relationships will depend in part on the amount and nature of the intervening space—how far it is from *A* to *B* and whether mountains, plains, seas, forests, farming, manufacturing, or dense or sparse populations lie between them. An atlas is designed to show as many of these things as possible in a series of maps—no one map can show them all without being illegible. Any map is thus an oversimplified and selective symbol of reality. Some distributions are best shown by a pattern of dots, some by lines, some by colors or shades, some by other

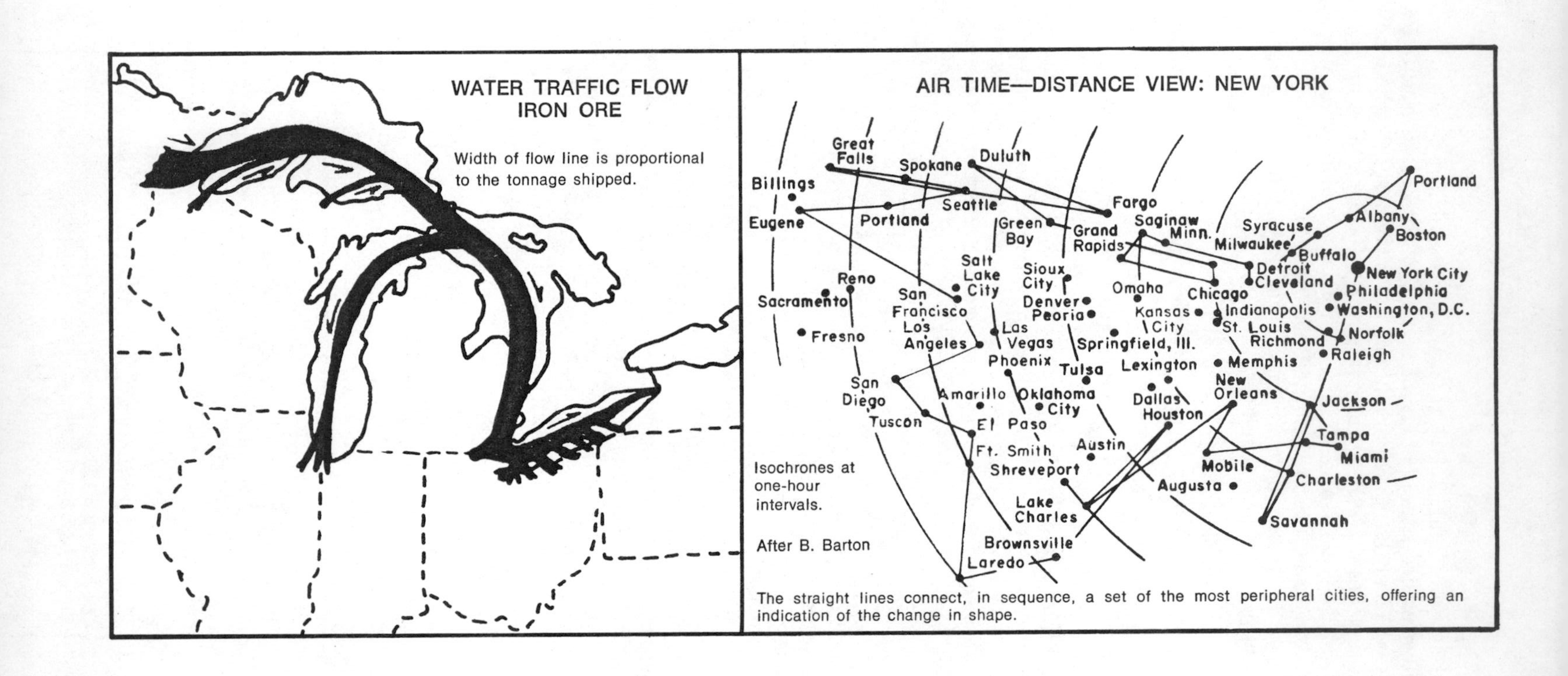
WATER TRAFFIC FLOW
IRON ORE
Width of flow line is proportional to the tonnage shipped.
AIR TIME—DISTANCE VIEW: NEW YORK
Great Falls
Spokane
Duluth
Billings
Seattle
Fargo
Portland
Eugene
Portland
Green Bay
Grand Rapids
Saginaw
Minn.
Syracuse
Albany
Boston
Milwaukee
Buffalo
Salt Lake City
Sioux City
Detroit
Cleveland
New York City
Reno
Omaha
Chicago
Philadelphia
Sacramento
San Francisco
Denver
Peoria
Kansas City
Indianapolis
Washington, D.C.
Los Angeles
Las Vegas
St. Louis
Richmond
Norfolk
Fresno
Springfield, Ill.
Raleigh
Phoenix
Tulsa
Lexington
Memphis
San Diego
Amarillo
Oklahoma City
Dallas
Houston
New Orleans
Jackson
Tuscon
El Paso
Tampa
Ft. Smith
Austin
Miami
Shreveport
Mobile
Augusta
Charleston
Isochrones at one-hour intervals.
Lake Charles
Savannah
After B. Barton
Brownsville
Laredo
The straight lines connect, in sequence, a set of the most peripheral cities, offering an indication of the change in shape.

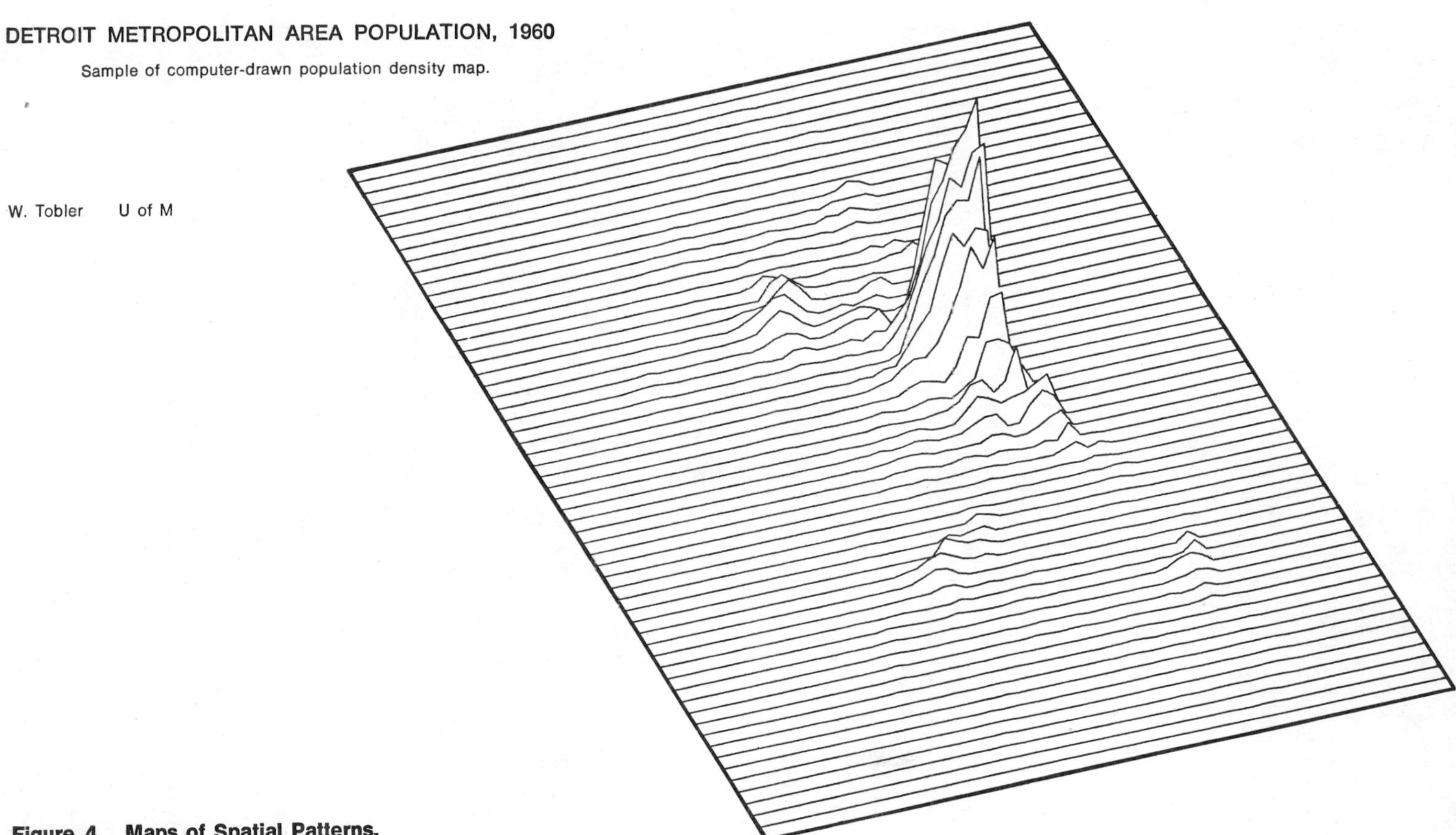

Figure 4. Maps of Spatial Patterns.

symbols. Relations may be indicated by lines such as transport routes, whose thickness varies with the volume of flow, or by the superimposition of distribution patterns to show coincidence. Routes or interconnections over space may simply be indicated by line symbols. Most relations, however, are more implicitly than explicitly indicated on maps, and they must be deduced or extracted.

A map is also a convenient inventory of selected information made available for immediate visual inspection, in the form of exact locations and spatial patterns which are manifestations of a variety of relationships. Such relationships may often be best or even first perceived through the process of making or interpreting a map and insights derived which were not guessed at earlier (see for example the discussion of China as a region in Chapter 6). The map is a powerful tool of analysis and not merely a collection of data. In addition to suggesting explanations and interrelations through its selective simplification of complex reality, it can direct attention to questions whose existence or importance were not seen before and can thus advance both theory and understanding. Anyone who uses maps should realize the varied dimensions of their usefulness and should look for more than simple locational data. Selected spatial patterns in cartographic form can raise and can suggest answers to a great variety of basic questions about the composition and interrelations of both the human and the physical landscape.

Scale[2]

Patterns of spatial form and spatial interaction can be examined at widely varying levels of detail. Map scale serves in effect as a filter, appropriate to the level of generalization required. It eliminates irrelevant or confusing detail and includes only what is directly helpful. A map is thus not only a selective and oversimplified symbol of reality, but a greatly reduced image. The degree of reduction is indicated by the scale of the map, one unit on the map representing a million units on the earth, for example, or a scale of 1 to 1,000,000. A map of the United States alone at this scale would nearly fill a small room, however, and maps of large countries or continents are therefore usually drawn at smaller scales like 1 to 20,000,000 or more, or of the world at 1 to 50,000,000 or less. ("Small" refers to the fractions; in these terms, 1 to 50,000,000 is a smaller scale than 1 to 1,000,000, although it may be used to show a much larger area.) Naturally, the smaller the scale, the less detail can be shown. A map with a large scale—like 1 to 62,500 (often used in the United States because it is an easily divisible fraction which is nearly the same as one inch to one mile or 1 to 63,360)—can show individual buildings, as in Map 1. For detailed analysis of small areas, such large-scale maps are essential, but most atlas maps are necessarily drawn at smaller scales. It will be helpful to examine the variety and progression of scales in an atlas and the bar scale com-

[2]Techniques for showing relief and some of the problems of scale are discussed in Chapter 12 on land forms.

MAP 1

PORTION OF KANSAS TO SHOW LARGE SCALE AND VERTICAL SCALE

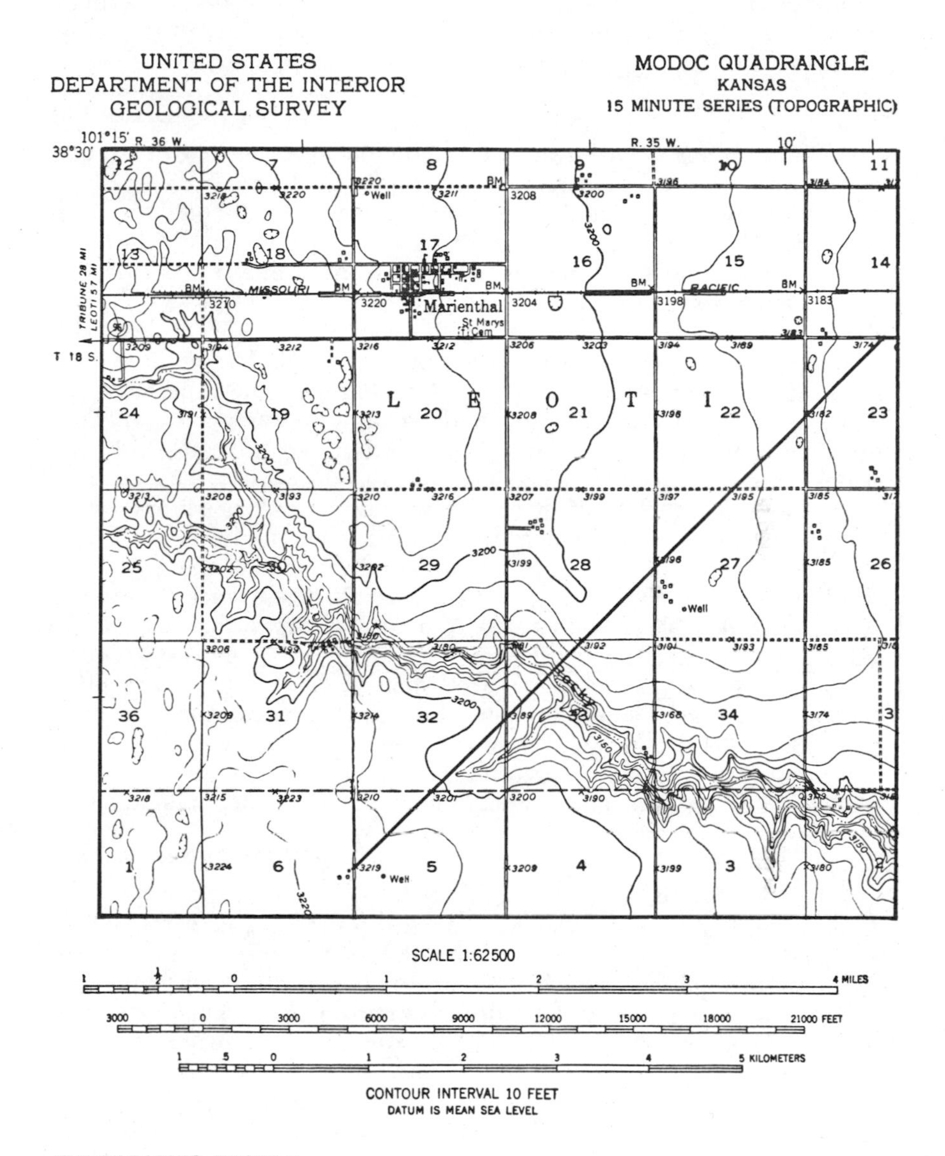

TOPOGRAPHIC PROFILE

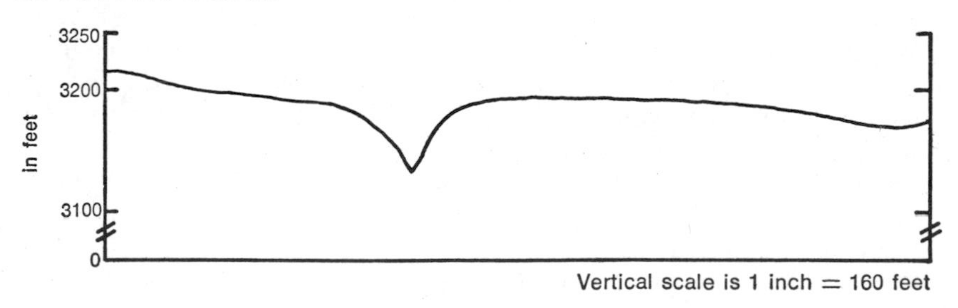

monly used; this is a line representing miles on the scale of the map, which is a great convenience in measuring distances, as in the sample in Figure 2.

Distortion

Despite their great advantages, maps must be used with caution because they are necessarily distorted representations of the earth as well as greatly reduced and oversimplified selective symbols. The earth is spherical and three-dimensional; a map is flat and two-dimensional. Long habit of looking at flat maps tends to make one forget that it is impossible to represent a spherical object or any part of its surface on a flat plane without distortion. The simplest proof of this is to lay flat a whole orange skin, or the two halves of a rubber ball. It cannot be done without tearing or stretching. A map distorts reality in shape, distance, direction, or total area. Most maps do mainly one or mainly another, depending on the method used to project the round earth onto flat paper. A few samples of projections and the distortions they produce are shown in Figure 5. Most atlases include some projections which distort shape and some which distort area, each useful for different purposes. Distortion is, of course, maximized in maps of the world as a whole (which cannot really be seen as a whole at all, since one cannot see around it), barely noticeable in maps of areas the size of the United States, and negligible for smaller areas. The distortion of small-scale maps also means that distances or directions are often shown inaccurately, and care must be used in measuring distances, especially on projections which do not show the true size of areas. The only approximately true representation of both shape and area is a globe, and flat maps should be used only with that model in mind.

Latitude and Longitude

For purposes of measuring and indicating exact locations, a grid of lines is superimposed on the map, as in Figure 6. The parallel lines extending east and west are called parallels of latitude, both north and south of the equator (which is midway between the north and south poles). These parallels of latitude are intersected by lines extending north and south, called meridians of longitude. Meridians are not parallel, because each of them originates and terminates at the poles; they therefore converge toward the poles and are most widely separated at the equator. One of these meridians of longitude is arbitrarily chosen as the base point of reference, or the prime meridian. For convenience and by established practice, the meridian which runs through the Royal Observatory at Greenwich near London is generally recognized as the prime meridian, although any other one could just as well be used provided everyone agreed on it. The fact that Greenwich has been generally accepted reflects the prominence of Britain in the modern world and the pioneer work of British scientists, explorers and map makers. Longitude is thus measured in terms of distance east or west of the prime meridian of Greenwich, and latitude in distance north or south of the equator. The earth is divided for these purposes into 90° of latitude and 180° of longitude, 90° north and south of the

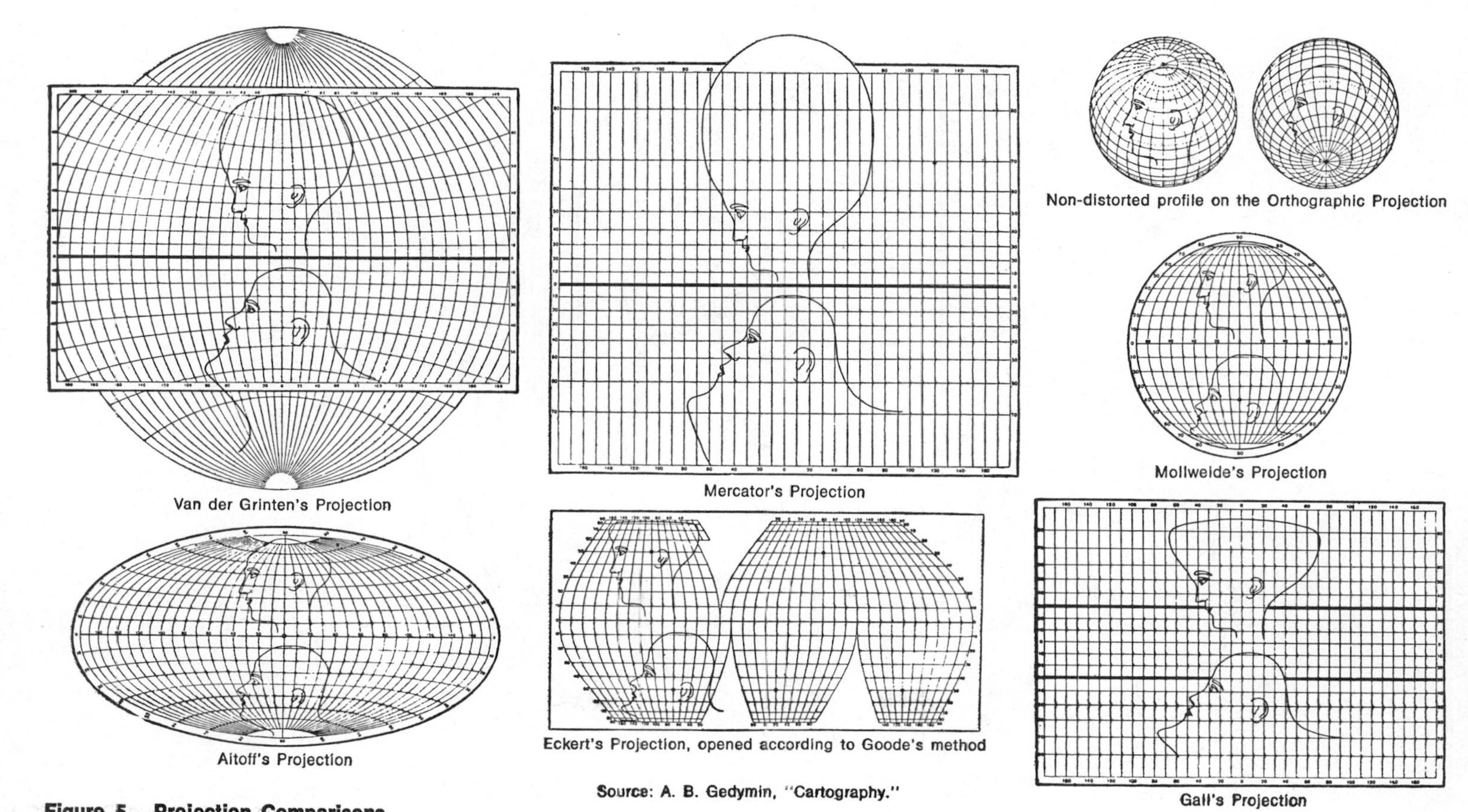

Source: A. B. Gedymin, "Cartography."

Figure 5. Projection Comparisons.

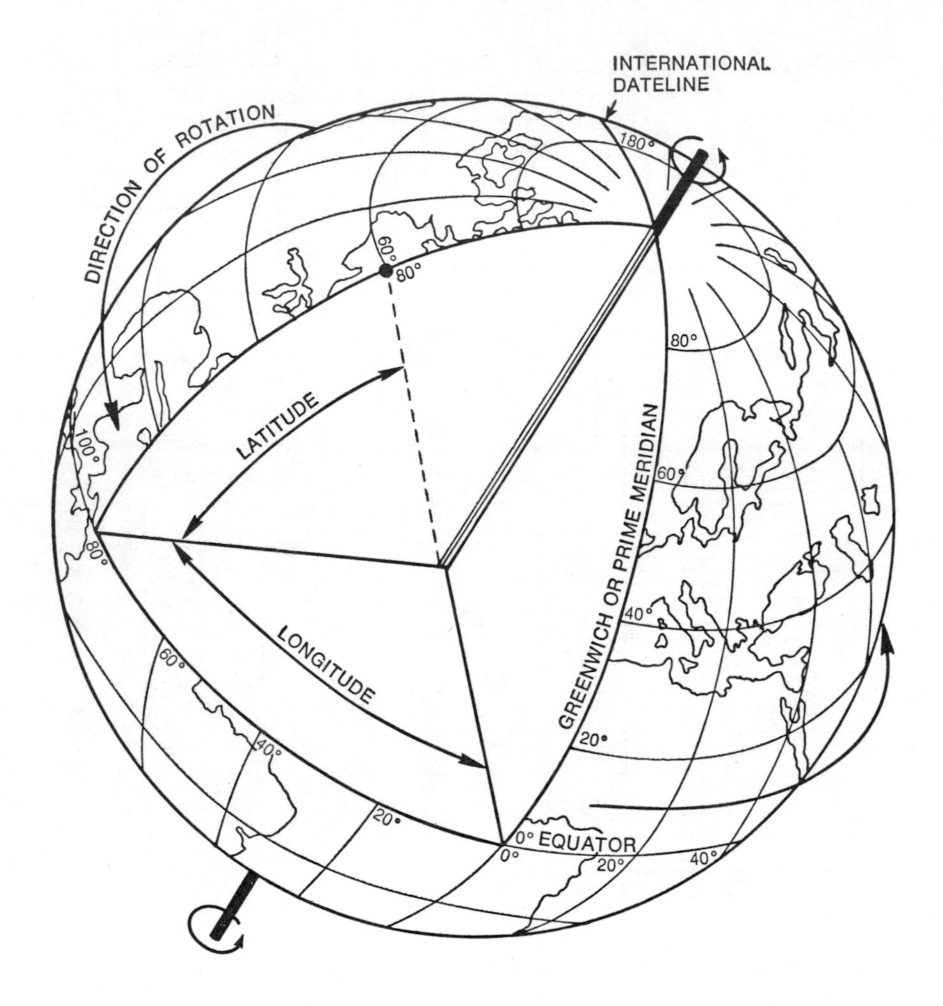

Figure 6. Latitude and Longitude. Cutaway globe that illustrates the angular relationships of meridians (longitude lines) and parallels (latitude lines). The point ● lies at 60° north latitude and 80° west longitude.

equator, and 180° east and west of Greenwich. The degrees are further subdivided into minutes and seconds for greater accuracy. Thus any location may be stated as so many degrees, minutes, and seconds north or south latitude and so many degrees, minutes, and seconds east or west longitude. By such means it is possible to fix locations very accurately and to pinpoint a house or a ship, for example, within a few yards.

180° east of Greenwich and 180° west of Greenwich are of course the same thing. Here another line must be drawn, in effect to separate east and west as the equator separates north and south, and also to mark the change in time from one day to another, in accordance with the rotation of the earth. Passing eastward from one time zone to another, clocks are advanced one hour, but passing eastward across the line of 180° of longitude, one enters the preceding day. Travel eastward, toward the rising sun, has progressed so far by 180° that it has caught up with and overtaken the day which arbitrarily began at 180° twenty-four hours earlier. Crossing 180° in a westerly direction, the reverse is true; clocks are set back from one time zone to another westward, but at 180° one enters the following day, which is arbitrarily beginning at 180°. The change of date at 180°, or the International Date Line, often confuses people; it is best to regard the day as beginning and ending at 180° so that the crossing of this line changes the date on the calendar.

Maps have been discussed in this chapter because they are an essential means to the study of space. Such study is only incidentally concerned with where individual things are and more importantly with how physical features, man, and his works are distributed over the earth, why these spatial patterns have developed, how they are interrelated, and how and with what effect spatial interaction takes place in all of man's activities. The map is both tool and symbol of geography's concern with spatial form and spatial relations.

3
Spatial Systems

GEOGRAPHIC ANALYSIS assumes that phenomena are distributed over the earth, or within any smaller part of it, according to one or more systems of arrangement which are at least potentially identifiable. This important aspect of geography is concerned with the spatial ordering of things and especially of human activity. Many of these spatial systems are more or less obvious: political units such as national states or counties, patterned systems of land use, such as rubber plantations or market garden areas around cities, the placing of houses along streets, and the arrangement of rooms and furniture within each house. Some are less obvious, such as the distribution of income, the incidence of particular diseases, or the spread of ideas. But whether obvious and directly observable or not, spatial patterns can be fully understood only when they can be delineated as individual or collective systems resulting from a combination of factors which can affect spatial behavior. Some patterns of spatial distribution are random or haphazard, but this situation is quite rare; nearly all distributions are non-random, and are shaped by particular factors, although the relevant factors may not always be easy to distinguish. The random-nonrandom distinction is, however, important to recognize in any study of spatial systems.

BEHAVIORAL SPACE

As suggested by the examples given above, spatial arrangement is evident in the entire range of human activities, beginning with a phenomenon which has come to be recognized as "personal space." Each individual orders his or her activities within or according to certain spatial limits and preferences. This ordering includes such things as the placing of furniture, the layout of a house, and even the ordering of the contents of a bureau drawer, a desk, or a kitchen cupboard. Some individuals are more sensitized than others to spatial arrangement;

some are said to have "a good sense of direction," and it is often these same people who order their personal space coherently. Others have a poor gross sense of direction and often have difficulty finding things even within the smaller and more personal scope of a house or a room where, from their point of view, spatial arrangement is not coherently ordered. Some people dislike crowds or small confined spaces; others enjoy and even seek such situations.

There are, however, patterns of personal and interpersonal space which tend to be consistent within any given culture, and to contrast with similar patterns in other cultures. One of the commonest of such patterns relates to interpersonal contact. In any given culture most people tend to accept or to require a certain minimum distance between themselves and others with whom they may converse; they also recognize and use a common set of gestures and other nonverbal signals often referred to as "body language." Each culture may have its own requirement of how large this interpersonal distance should be and may have its own, quite different and accepted vocabulary of "body language." This concept is related to what is sometimes rather misleadingly referred to as "territoriality," a concept derived from the study of behavior of other animals where the notion is perfectly appropriate even though it may not fit human behavior as well. Most species of animals and birds are dependent for their survival on the control or exploitation of defined pieces of territory which can provide them with food or ensure their ability to reproduce; they therefore attempt to defend that territory against trespass or use by others of the same or other species and may also stake claims to it by various means or signals. Bird songs, for example, are at least in part a means of establishing claims to a given foraging and nesting territory. Man's survival problems are different, primarily because he has passed beyond the hunting-and-gathering stage and is able to assemble what he needs over great distances. Nevertheless, there is an element of territoriality still involved in individual perceptions of personal space, and, by successive extension, in perceptions of family, home, or group space as well as of national space. Perhaps the major difference between man and other animal species in this respect is that animals have long ago learned how to order potential conflicts over spatial arrangements and allocations, whereas man still engages in both individual and mass destruction in pursuit of a territoriality which is not adequately provided for or regulated.

SPATIAL PERCEPTION

Each culture has its biases, and its own idiosyncratic perceptions of space. Most individuals, and most culture groups, have a spatial perception which is both warped and incomplete. You have probably seen examples of cartoon maps, or cartograms, made to illustrate this point, for example, the New Yorker's view of the United States or the American tourist's view of Europe. In these the viewer recognizes the existence, location, importance, or size of something in accordance with his perception of what is important. Individuals have a similarly warped and incomplete mental map even of the much smaller space which they occupy or move around in. Reality is immense, and what it is possible to notice, remember, and organize is only a small selection, based on what is considered important. On a daily journey to work along the same route, a commuter will miss (that is, he will sort out of his conscious attention) most of what is directly perceivable, and will order that particular space corridor in his mind in accordance with his own criteria, which, for example, may well reflect travel time rather than distance. He or she may be totally ignorant of what lies immediately outside the space corridor through which the daily journey to work moves but may be familiar in detail with other more remote areas or have an accurate impression of their location and nature even without having gone there. Culture affects individual perceptions of space in consistent ways because it tends to define what is important, and is also the principal medium through which individuals learn or acquire information. North American culture is relatively self-contained, isolated, and self-satisfied. North Americans are most closely aware of western Europe, but even their information about Europe is scanty and often inaccurate. Much of the rest of the world is perceived even more vaguely, if at all, as many sample studies and questionnaires have demonstrated.

Such cultural biases are perfectly natural and universal. However, as travel time shrinks and all parts of the world become increasingly interdependent, or as conflict threatens as a result of imperfect knowledge or understanding, warped and incomplete perceptions of global reality become dangerous. There is less excuse for geographic ignorance on the part of North Americans, or for the dangerous biases which their culture tends to produce, than for similar attitudes on the part of less wealthy populations in the past. China is an example of a culture whose biases resulted in a warped and incomplete perception of global geography which left it poorly prepared to cope with the pressures of an expanding West in the nineteenth century. China was rich, powerful, and self-satisfied, effectively isolated from any extensive contact with other advanced cultures by the mountains, rainforests, and deserts which surround it along its landward frontiers, and was largely ignorant of the world beyond East Asia. The "friction of distance" which slowed or prevented effective contact across the thousands of miles of Eurasian desert or empty Pacific helped to convince the Chinese that the rest of the world was of little consequence. Westerners in particular were looked down on when they turned up in small numbers on the China coast in the eighteenth and nineteenth centuries and were dismissed as "backward." The Chinese made few distinctions among them and refused to take seriously people who seemed to them simply savages, coming from places so far away that the Chinese had never heard of them and had no way of fixing their location, except to gather that people who lived so far beyond the Chinese sphere must necessarily be backward, since they could not have had the advantage of contact with Chinese civilization.

George III of England sent an emissary to the Emperor of China in 1793 in an effort to promote trade and diplomatic exchange. His requests were refused, for as the Emperor's reply put it, the Celestial Empire "enjoys all things in abundance" and had no need for trade with barbarians. King George's ambassador was designated as a representative of a "barbarian chieftain" who had come to China bearing tribute to the Emperor, and the Emperor's letter to King George (which must have infuriated that quick-tempered monarch) stressed the same theme: "I note with pleasure your respectful spirit of submission. I do not forget the lonely remoteness of your island, cut off from the world by intervening wastes of sea." The English perception of spatial arrangement was of course exact, and, at least for the ensuing century or so, more in accord with reality. Less than fifty years after this exchange, British military forces easily defeated the Chinese, occupied several major Chinese cities, and imposed a humiliating settlement which was to set the pattern for the next hundred years. The diagram in Figure 7 is an effort to illustrate traditional Chinese perceptions of space, and the degree to which their picture was inaccurate or incomplete.

The line represents the level of effectiveness or power. The solid sector shows the part of reality which the Chinese knew, the area immediately around them, including neighboring cultures which were in fact less accomplished or powerful than their own as of about 1800. The dashed sector continues this line beyond the limits of Chinese knowledge and shows in some approximation that in fact distant areas to the west and east were rapidly becoming as powerful as China, or more so.

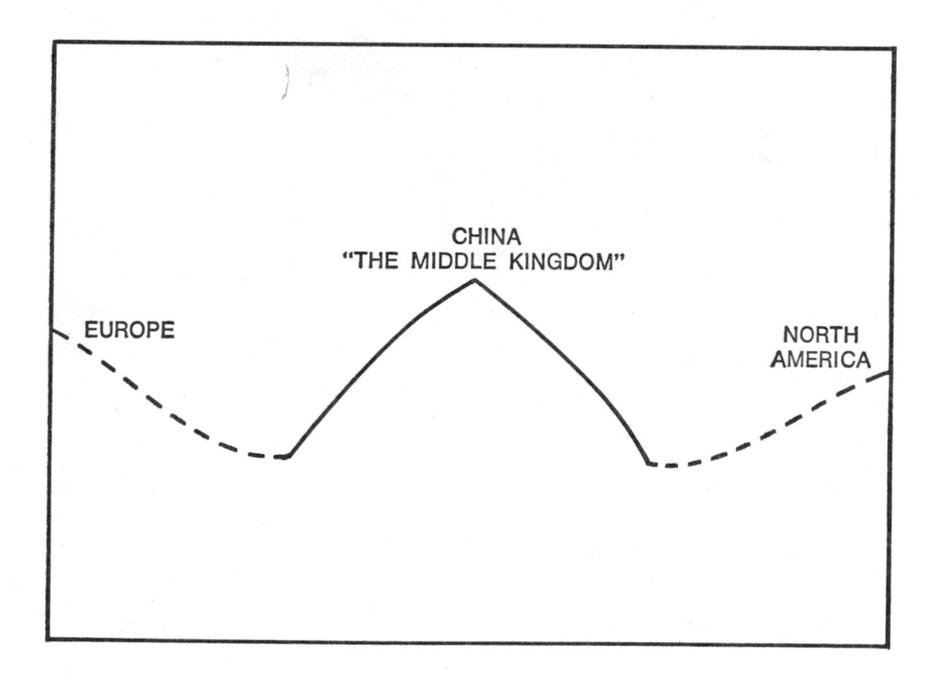

Figure 7. The Chinese World About 1800.

MAP 2

GROWTH OF THE DETROIT URBANIZED AREA

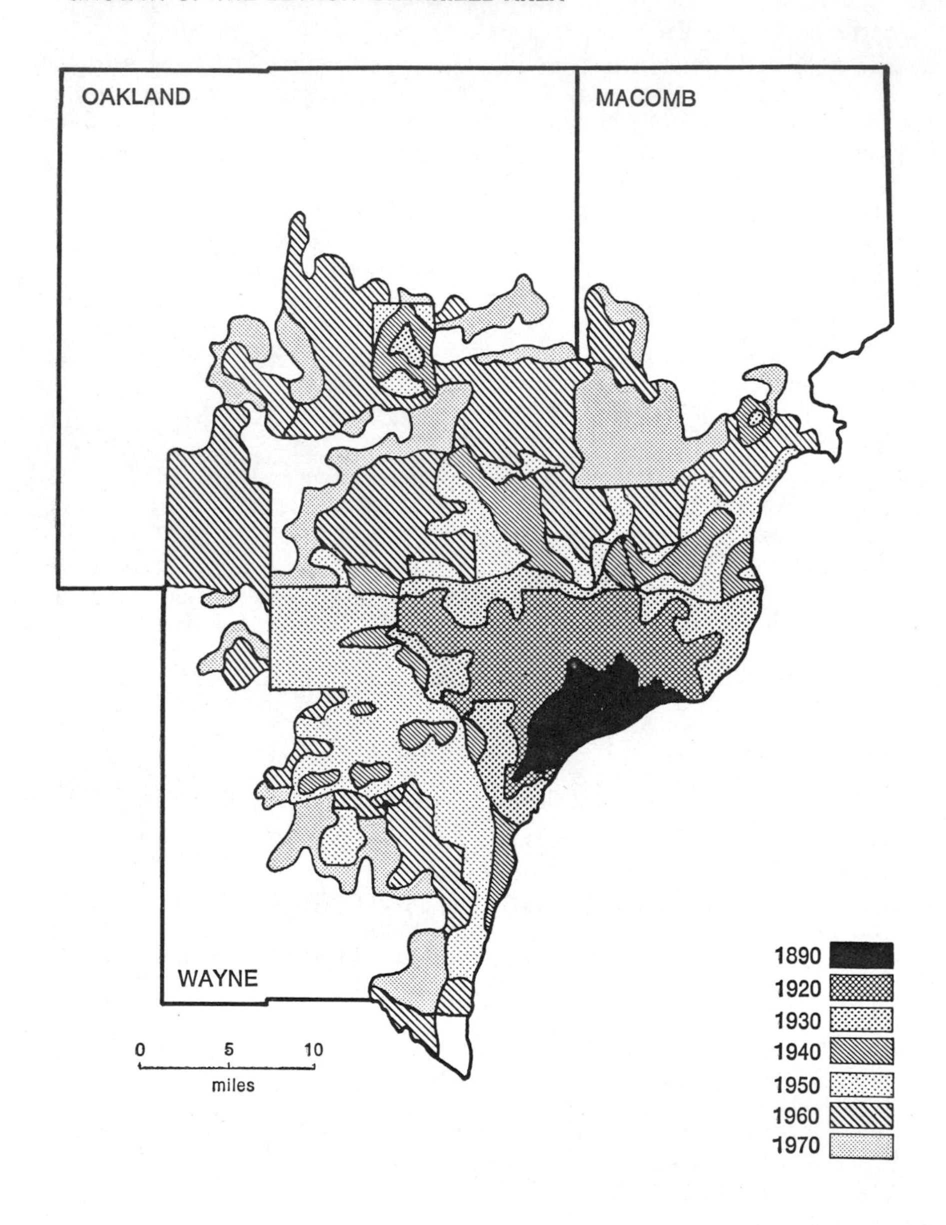

Sources: Judy S. Humphrey, Segregation and Integration, Advancement Press of America, 1972. Donald R. Deskins, Settlement Patterns for the Detroit Metropolitan Area, 1963.

The Chinese had to learn this fact the hard way and were ultimately obliged to change not only their perceptions but important aspects of their own culture in order to survive in the modern world.

LIVING SPACE

Perceptions of space vary, over time, from place to place or culture to culture, and among individuals. But concrete phenomena and actual flows are arranged in space nonrandomly and according to identifiable systems; it is part of the geographer's job to identify and analyze them. Settlement patterns make a good illustration, and are perhaps the most widely studied forms of spatial systems. Cities, towns, and villages grow up at certain locations and perform certain functions primarily as a result of spatial interaction, as discussed in the preceding chapter. The nature and amount of the settlement's interaction with other places will largely determine its size and its own nature and will even influence its internal spatial arrangement. Modern Western cities have exploded in size during the past century in particular, as the means of transport have multiplied and it has become possible to move larger amounts of goods more cheaply over greater distances and at greater speeds. This enables the city to perform more functions on behalf of larger areas, to bring in the food and other supplies it needs to maintain its own population, and also to drawn its own work force from greater distances so that the urbanized area (including commuter suburbs) has rapidly extended as cars, buses, and trains bring larger areas within bearable travel time from the city.

The arrangement of agricultural villages reflects the same kinds of spatial constraints or circumstances, even in areas were mechanized transport is absent. Where movement is by foot, animal power, or water, the spacing of large settlements will reflect most importantly corridors along which bulk goods can be moved cheaply, primarily waterways or sea coasts. Smaller settlements tend to be spaced in accordance with local agricultural productivity and to be closest to each other where productivity is highest. This pattern is seen in most of monsoon Asia, for example, where the distribution of villages on an agricultural plain looks like the sketch in Figure 8, based on an aerial photograph of a small part of south India.

In these situations, productivity per acre is high and fields are small. They are grouped directly around the village, and the farm workers from each household are easily able to walk to their plots every day. The size of the village is effectively limited by the number of households which the area can support on the basis of agricultural land within easy walking distance, a measure which is of course determined also by the productivity of the land. As shown in Figure 8 an agriculturally productive landscape tends in this way to be broken up into groups of small fields clustered around villages that are relatively close together, with the outer borders of the fields of one village adjacent to those of the next village. In a less productive setting, village spacing would be wider. Like most spatial systems, these patterns

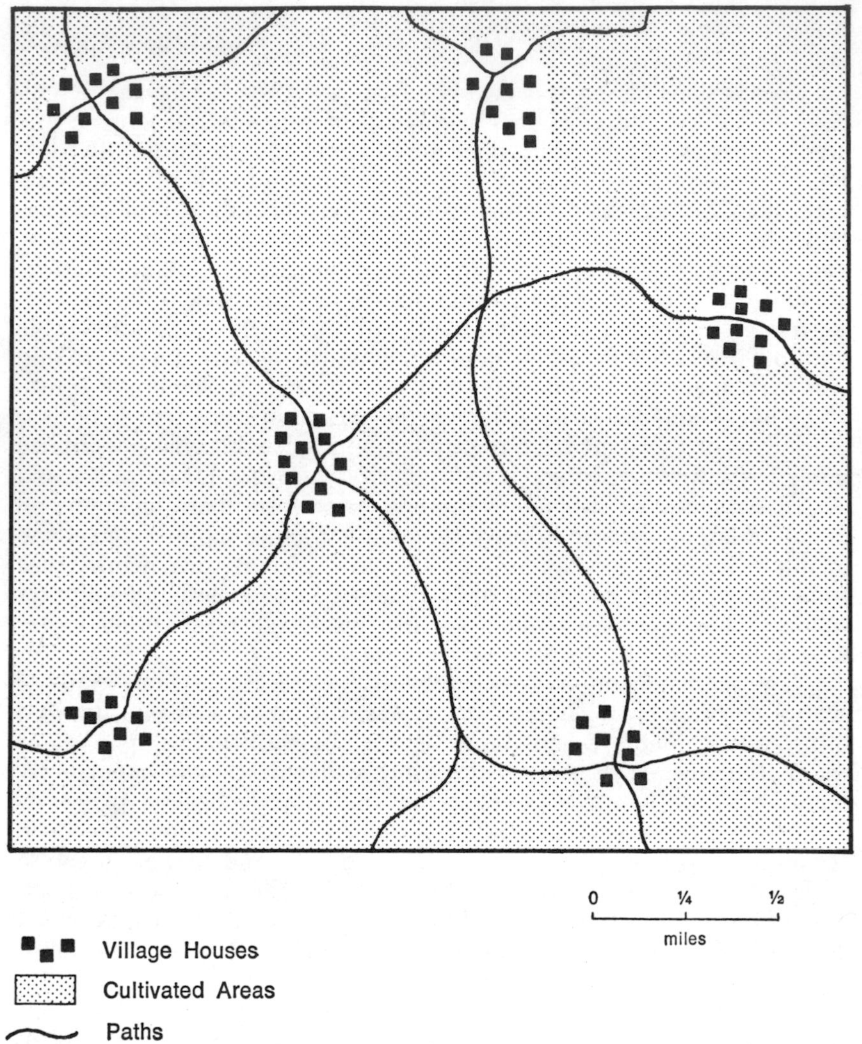

Figure 8. Village and Agricultural Landscape on a South Indian Plain.

reflect also the principle of least effort, whereby things are arranged so as to minimize movement, labor, or expense. The same principle can be seen in the spatial layout of modern cities (see Maps 2, 3, 4, and 5, for example, and Figure 9), and in the arrangement of agricultural land uses at varying distances from a city, as discussed and illustrated in Chapter 10, Pages 159–180.

Spatial arrangements within each village reflect their function, and their spatial constraints, as shown in the sketch in Figure 9.

Houses are closely clustered along a central or single path, and fields radiate out from the village on all sides. In such a situation, people necessarily live closely together, and various forms of cooperative effort are involved. Relations with other places and other people tend to be minimal, but within the village relations and interdependence are close. These circumstances are in turn reflected in the traditional social structure characteristic of monsoon Asia: an exceptionally closely knit family system, a firm web of intra-village connections, little scope for individual as opposed to group interests or values, and a generally closed or self-protective face to the extra-village world. All of these characteristics have clearly been influenced importantly by the spatial system, the particular frame in which village Asia existed and to a large extent still does exist.

The contrasting spatial circumstances of the modern Western city are clearly related to its contrasting social and cultural characteristics, and to many of its contemporary problems: individualism, alienation, rootlessness, mobility, and rapid change. Modern Western cities are also primarily commercial phenomena, and their internal spatial patterns and forms are shaped to a major extent by their commercial/industrial functions and by the commercial value of urban land. The layout and architecture contrast with those of cities whose chief functions are ceremonial, symbolic, or administrative, including most cities built before the nineteenth century (see the reading list at the end of this chapter).[1] In most contemporary American cities, land values tend to reach a peak in or close to the city's central business district but to fall off away from the center in an irregular pattern. Large space-users (railways, most factories) must obviously seek cheaper land, but functions whose success depends on easy access in high volume, especially department stores and other retailing, must seek the center or outlying centers where transport access is maximized. They tend to cluster near points where pedestrian traffic is greatest or where individual customers can most easily reach the area by car or by public transportation.

With the advent of the private car and the consequent decay of public transportation in many cities, department stores and other retailing, formerly concentrated largely at the center where public transport lines converged but where the private car was at a disadvantage, have had to move or to establish stores in outlying locations

[1]An excellent sample of the literature on this subject is Paul Wheatley's *Pivot of the Four Quarters* (Chicago, 1971); Chapters 3, 4, and 5 of Wheatley's book deal in great detail with early cities in Asia, Europe, Africa, and the Western Hemisphere as cosmo-symbolic ceremonial centers: their morphology, functions, and nature.

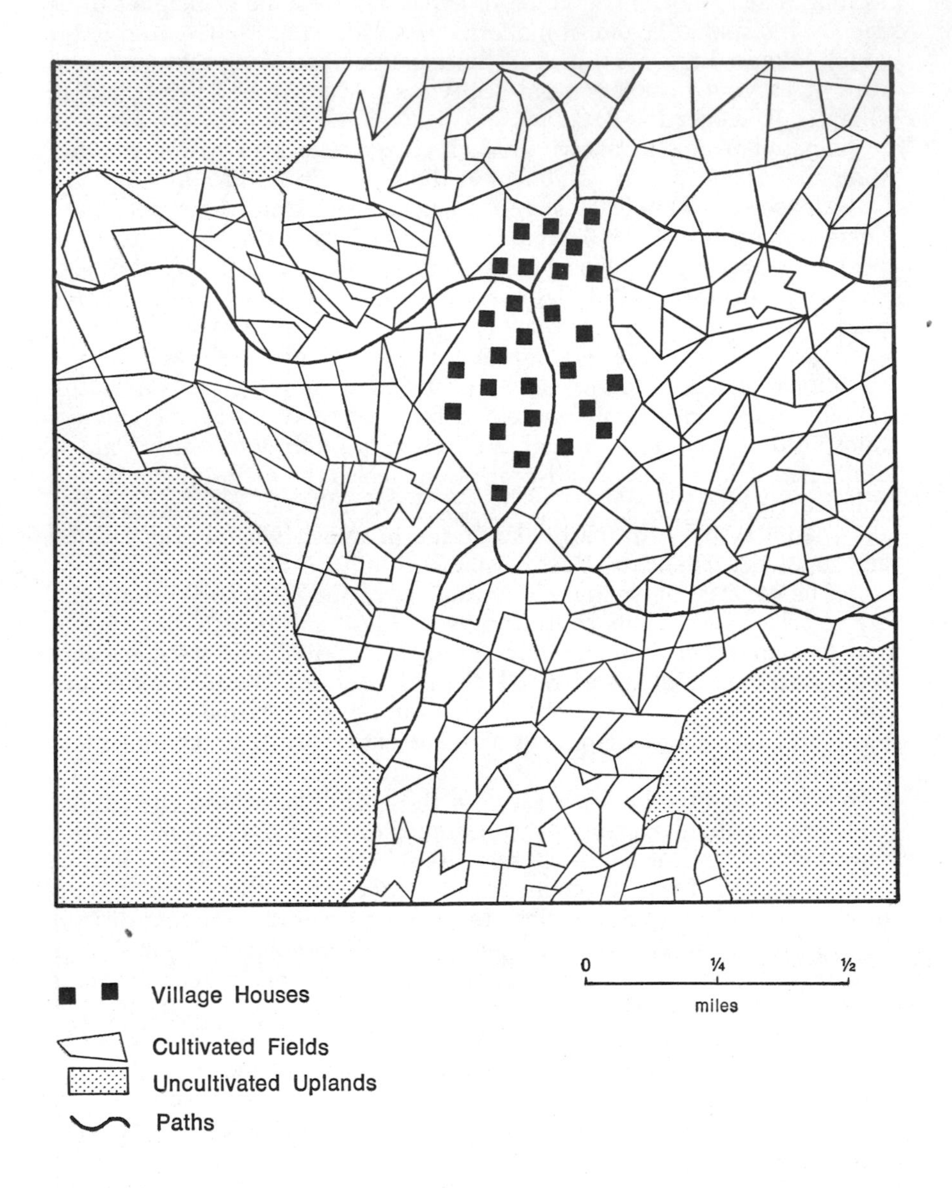

Figure 9. Standard Asian Village Pattern.

where car access is better and where ample lower-cost space is available for huge parking lots, many times the size of the store area itself. Many of these factors and changes are reflected in Map 3 of urban land values and associated land uses.

It is obvious that land values are reflected also in the pattern of residence, including its segregation by race, cultural origin, and occupation, since these factors are related to income. Black residential patterns make the clearest case because, at least in most American cities, Blacks are the largest easily identified separate racial and cultural group which also is largely restricted to low income occupations. Map 4 shows Black residence superimposed on the previous map of urban land values and land uses. Boston's Chinatown is shown in Map 5.

SPATIAL ASSOCIATION AND MIGRATION

However, land values and occupation/income are not the only factors which affect residential clustering by racial or cultural groups in most cities. Much depends on the less easily measured influence of cultural ties. Culture varies, of course, even within a single national group or single city; it is affected, for example, by occupation, by income, by sex, by age, and by other similar factors which also help to form interest groups sharing certain attitudes, values, or circumstances. Members of a suburban bridge club or tennis club, for example, may have relatively little in common and minimal interaction with a central city factory workers' bowling club and will have different points of view on most subjects. Even their speech will differ to some degree, as well as their preferred diet, dress, and musical tastes. Culture may indeed be as important, or more important, than race in influencing people's choices of interpersonal association and interaction, which in turn help to determine residence patterns. Clearly discernible racial differences, such as skin color, do unfortunately greatly sharpen distinctions in the minds of many people and lead to (often unfounded) stereotypes about the supposed behavior of supposed racial groups. Perception of differences may lead to segregation imposed by the dominant racial or cultural group, as is regrettably the pattern in most American cities, but it may also tend to strengthen the sense of group solidarity on the part of those who are discriminated against. They may have in common not only a set of cultural traits and thus be inclined to seek one another's company but may also share common oppression. Both sets of factors tend to affect residence and interaction patterns.

Such voluntary association may especially appeal to groups or individuals who are relatively recent arrivals in a city or a national unit. As strangers who also suffer to some degree from linguistic, other cultural, or economic handicaps in competing or interacting with the existing dominant group and who may in addition be discriminated against, they understandably tend to stick together. This pattern has been seen, for example, with Asians in the United States and with Black migrants to northern cities from the south, both of whom have tended to settle in quite sharply segregated patterns associated with low-value (but often high-rent) areas close to the center of the city

MAP 3

LAND VALUES AND LAND USE: CHICAGO, 1960

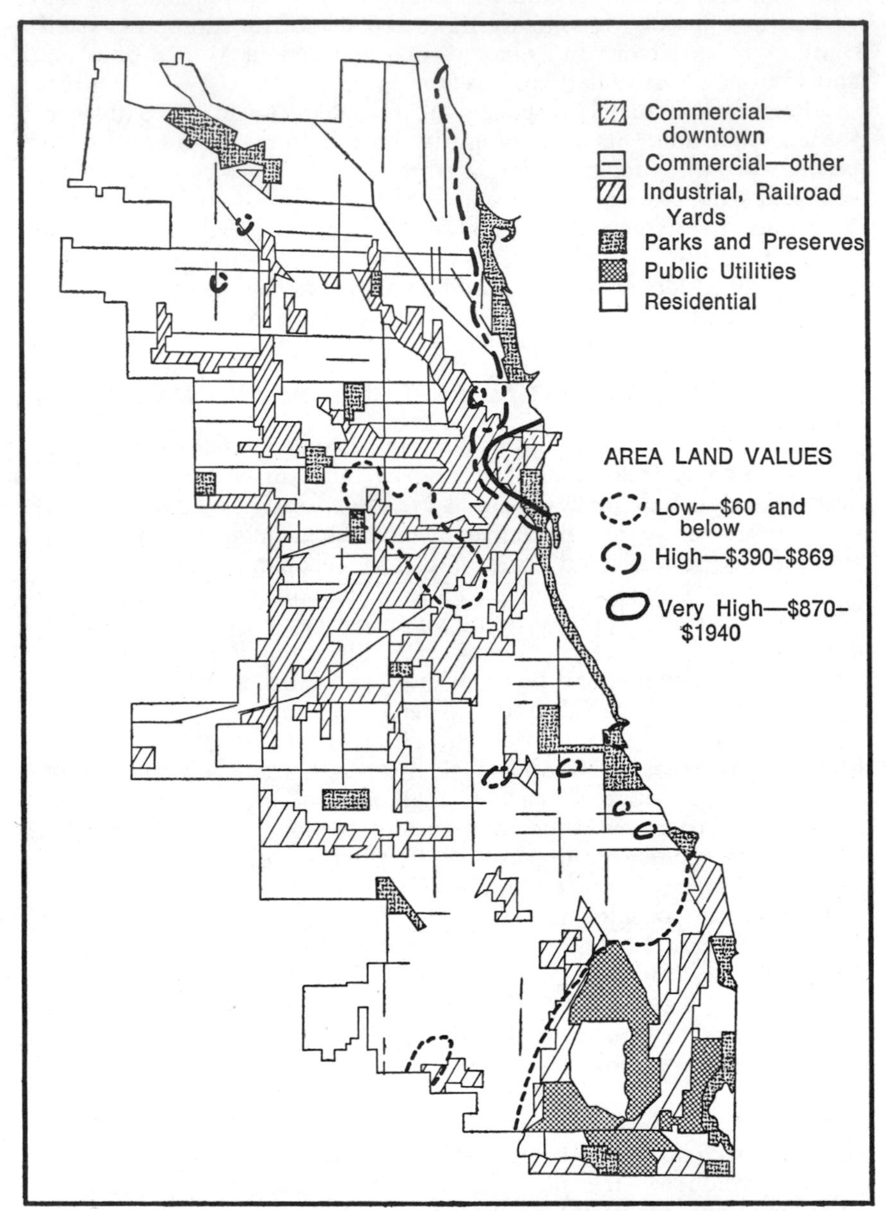

Sources: 1. *An Atlas of Chicago's People, Jobs and Homes,* Community Renewal Program, 211 West Wacker Dr., Chicago 6, Ill. June, 1963.
2. Maurice H. Yeates, "Some Factors Affecting the Spatial Distribution of Chicago's Land Values, 19101960," *Economic Geography,* Vol. 41, No. 1, Jan. 1965, p. 61.

It is difficult to show three different kinds of things (two of them expressed in a number of categories) on one map without sacrificing some clarity. To somewhat resolve this cartographic problem, it has been necessary to simplify complex reality by abstraction or generalization into a limited number of categories that, therefore, give an incomplete picture of the actual variety of land uses and land values. Note, however, how high land values hug the lake shore for both residential and commercial uses (the highest values of all are in the commercial core of the Loop, the retail, financial, public building, and office center of the downtown area). Black residence is associated

MAP 4

BLACK RESIDENCE ADDED TO LAND VALUES AND LAND USE: CHICAGO, 1960

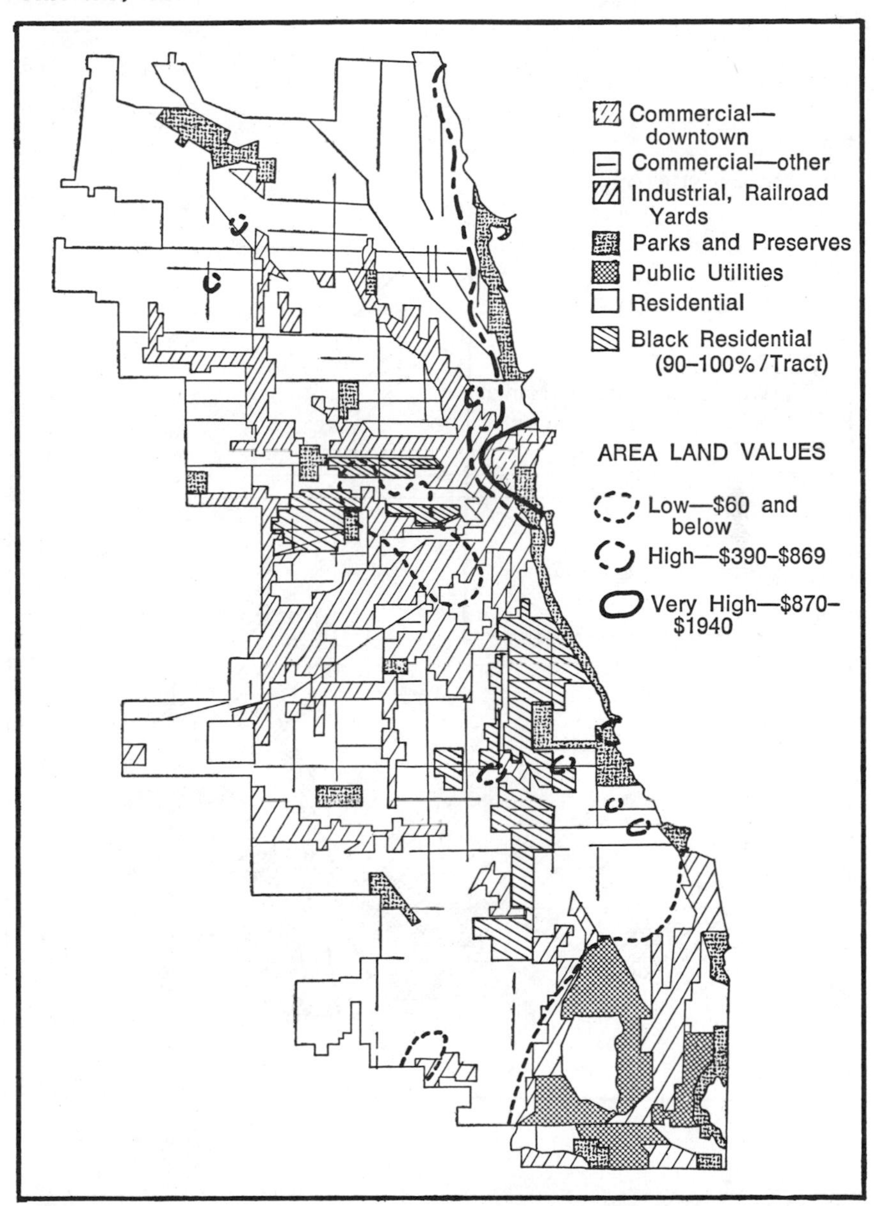

with several of the lowest value areas and with areas where industry makes residence less desirable, but this is far from a perfect fit. Part of the problem results from over-generalization of the categories; the fit would be closer if more detailed breakdown were possible. But not all poor people are Black, nor are all Blacks poor. In addition, Blacks and other disadvantaged minority groups often, ironically enough, occupy relatively high-rent structures, and investment in such buildings may be very profitable for those with capital, which tends to mean relatively high land values. This does not, of course, mean that buildings or living conditions in them are good. Such patterns are the result of racial segregation and exploitation that leave landlords more powerful and freer in their choices than Black tenants (or poor white tenants).

Many other aspects and implications of spatial patterns in contemporary American cities may be detected or deduced from a study of Maps 3 and 4, which provide a further illustration of the uses of cartographic expression of data.

MAP 5

CHINATOWN: BOSTON

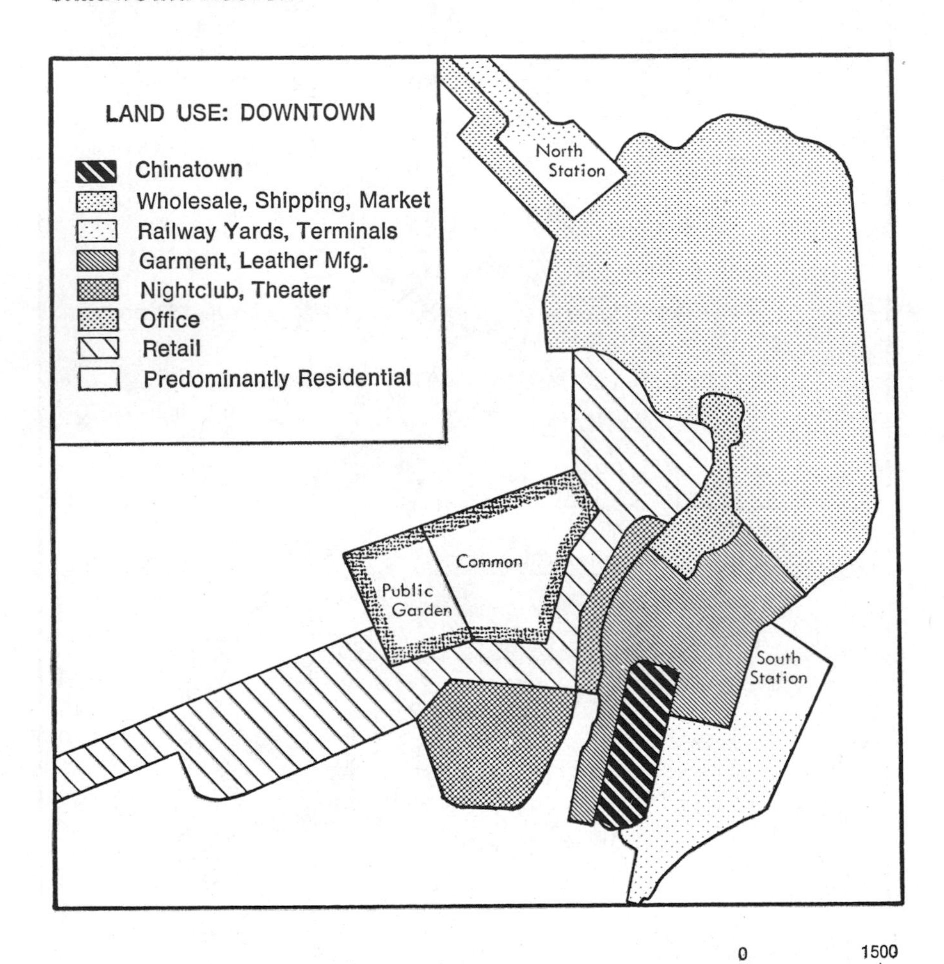

where some public transport is available but where manufacturing, wholesaling, or the more general urban blight which has grown around the central business core of most American cities make these areas undesirable to more advantaged groups and hence available to others at the bottom of the heap.

The study of migration patterns: what kinds of moves have taken place, which groups and individuals have been involved and over what routes, and what kinds of factors affect this process, is a rapidly developing field of analysis in which many geographers are involved, since it is an obvious aspect of the spatial dimension of human society. Migration affects, of course, not only the migrants themselves but the areas to which they move, and those which they have left, whose character also changes, in some cases fundamentally. Most migration takes place under some form of duress or pressure, usually economic need, unfortunately often also discrimination or more severe oppression. Both may however be affected by individual perception, economic need necessarily so since it is of course a relative matter. Before the individual can rationally decide to move, he or she must conclude that circumstances elsewhere are in some way better. His or her information about other places may be and usually is imperfect or even erroneous, and for most people migration is a painful and difficult enterprise not undertaken lightly.

The distribution of population is of course itself a spatial pattern, and is discussed as such in Chapters 7, 9, 10, 11, 12, and 13 as it is related to the different elements of the natural environment which together form the basis on which man lives. Population is discussed in greater detail in Chapter 13, in its relationship to resources and resource use, and in terms of the problems attendant on the explosive increase in human numbers in the past century, an increase still continuing. However, population pressure has long been a direct factor in migration patterns, or an indirect one as it has contributed to wars. In general, most of the great migrations of the past, and through perhaps the first quarter of the twentieth century, have been in response to economic pressures, especially famine, or to wars. Such movements are still from time to time taking place—for example in Bangladesh or in Indochina—but especially in the so-called developed or affluent world most migration has become increasingly a matter of individual choice and often takes place within a given national unit, in search of still greater economic opportunity, or of amenities such as milder climates.

DIFFUSION

Most potential migrants base their decision on information about other places derived from word-of-mouth reports from other migrants; these reports are usually diffused through a chain of intermediaries and hence may become distorted or may be unrepresentative in the first place. The whole process of information diffusion, an important dimension of every aspect of change or innovation in human behavior, takes place over space and hence is properly a subject for geographical analysis, as is the diffusion of any change. Practically all aspects of technology and culture and most plants, animals, and diseases were once confined to relatively small areas or were initiated in a particular location and have since spread more widely or globally.

The timing and the means by which these patterns of diffusion take place have become one of the fastest growing fields of investigation in the past twenty years within the discipline of geography, concerned as it is with the spatial dimension. Human migration is merely one form of spatial diffusion, but one which has been happening since the emergence of Homo sapiens. The dissemination of crop plants, such as wheat or potatoes, and the unplanned spread of wild plants as unintended human baggage may be almost as old and is still continuing. In most areas of the world, relatively few of the cultivated plants or of the wild species ("weeds") which invade fields, cleared areas, and roadsides are originally native to that area and many have been introduced very recently; the same applies to a surprising number of wild as well as domesticated animal species. Perhaps the clearest illustration is provided by pheasants and chicory (a common roadside or field "weed" with lovely light blue flowers) in North America. Both of these are now almost universally diffused in this continent, but both were introduced here originally in very small numbers in the second half of the nineteenth century. Two of our commonest birds, the English sparrow and the starling, were also introduced, in a few pairs each, in the 1870's. Horses, cows, pigs, sheep, and chickens were all introduced into the Western Hemisphere after 1492, as they had been into Europe (from Asia) some millennia earlier. With them, and with the cultivated plants introduced, came animal and plant diseases, as human diseases spread with human migration.

With the greatly increased human interaction and actual movement of peoples and goods attendant on improvements in transport in the past few centuries, human diseases have been able to spread much more widely and rapidly. Diseases seem also to follow a cyclical pattern of incidence and virulence, pulsing in spasmodic waves over time and, with the newly increased means of diffusion, over space. The frightening increase of human population has also created better conditions for the spread of disease, through crowding, through eroding the nutritional or other economic bases of subsistence, and simply by providing more hosts. In time, most populations acquire a degree of immunity or at least resistance to many diseases which are endemic in the area they occupy. New diseases, imported from another area, may be totally devastating, as has happened repeatedly in the course of human migration. The common cold, for example, was a dangerous and occasionally fatal illness to some North American Indian and Eskimo groups when it entered the continent with early European settlers; measles, which they also brought with them, brought death to the great majority of the local population who contracted it. Only widespread intermarriage (or interbreeding), which in effect transferred European genes to the local population and hence diffused a degree of resistance to these and other imported diseases, prevented an even greater decimation.

Most attention has been paid in the rapidly developing study of diffusion to the spread of technological and cultural innovation, which is indeed a fascinating process, in part because one can see it operating and because it is continuing to take place in new forms on a short time scale. The saga of the hula hoop makes a good example, with its roughly five-year span during which it seemed to appear and to spread almost overnight and then equally suddenly to disappear except for a few die-hards. Like most innovations, however, it was not universally diffused, and in this case tended to be rather sharply confined by age and only slightly less sharply by sex. As far as I know, no one has made a systematic study of this phenomenon, perhaps it is considered too trivial, but it involves most of the questions which underlie any aspect of innovation and its diffusion. The same applies to the somewhat similar and so far more persistent innovation of the frisbie, which having largely conquered North America (although here too on a restricted age and sex basis) is now making inroads in Europe and even in small parts of Asia, Africa, and Latin America.

Systematic study of the contemporary or recent diffusion of innovation has so far dealt with a variety of perhaps less trivial examples, including ownership of radio or television sets, telephones, and automobiles, the spread of new crops and of new cultivation techniques or machines, and the changing spatial pattern of attitudes (political, ecological, racial, religious), clothing styles, dietary innovations (for example, pizza), drive-in banks, and saunas, to name only a few examples. Each of these innovations tends to depend for its spread on a somewhat different chain of circumstances and a different network of diffusion, including of course the means by which information about the innovation is made available to potential adopters and the degree to which potential adopters are influenced by information or example from particular sources. Each innovation is likely to have its own discrete pool of potential adopters, many of them not necessarily susceptible to other forms of innovation, or in any case less likely to be in touch with a different interpersonal or intergroup network which might stimulate them in favor of another innovation. Physical proximity does of course assist the spread of innovation from one adopter to another, but commonality of interest or circumstances may be equally or more important. Wheat farmers, for example, are more likely to be directly or indirectly in touch with and to adopt innovations from one another or from a common network such as an extension service, even though they may live considerable distances apart, than they are from their immediate pig-raising or school-teaching neighbors. Other more general innovations, such as telephones, are likely to spread through a series of discrete networks and on a different time schedule: first perhaps within a given income or occupational group in a city, then within similar groups in other cities, then to other urban groups, and finally to a series of rural groups, each constituting a separate network; or the innovation might diffuse primarily through income or occupational networks regardless of urban/rural distinctions.

SPATIAL INEQUITIES

Whatever the changes and tendencies toward uniformity brought about by the spatial diffusion of innovation however, the world remains enormously varied, both in its regional range of physical diversity and in the way in which people live. This is not merely a matter of regional culture, preferences, and attitudes but involves gross disparities and inequities in the material bases of life. Two-thirds or more of the world's people, sometimes euphemistically called the "developing" world, live and die close to and periodically below the margin of subsistence, while the remaining one-third enjoy ever-increasing affluence. Man has not yet learned how to diffuse the economic abundance which his new technology makes possible. Most of that new technology is used instead to strengthen the power and wealth of those who already have it and to support international rivalries and conflicts aimed at the same objective. Relatively speaking, the poor grow poorer and the rich richer as the gap between the "developed" and "developing" sectors continues to widen. These inequities are overlaid onto a pre-existing pattern of racial and cultural differences. A series of circumstances of very recent origin has during the past two centuries—a brief moment after all in the course of human history—brought wealth and power mainly to the people of Europe and North America (plus Japan), most of whom share a common racial and cultural grouping; the rich of the world are predominantly white. It is not easy for the rest of the world to avoid resentment, and to use pre-existing racial, religious, linguistic, and other cultural differences as symbols of rivalry or as divisive rallying points.

These kinds of spatial patterns and the factors which account for them are clearly the most important of all. The geographer's efforts to identify and analyze the spatial dimensions of human society need to be concentrated on those which help to divide mankind in these destructive ways, and on the search for answers which can help to ensure that differences, which will always exist, do not lead to injustice or to conflict. We are beginning to learn a good deal about spatial patterns of difference and interaction. That developing knowledge needs to be used.

Questions for Further Study and Discussion (Chapters 2 and 3)

1. What is the relation between specialization, exchange, and economic development?

2. In what ways is urbanization dependent on spatial interaction?

3. How does spatial interaction affect regional distinctions? How does it affect political boundaries? How do regional differences and political boundaries affect spatial interaction?

4. Select an example of spatial interaction within your city or town, within your state, and within the United States, and analyze each.

5. Are there any aspects or characteristics of human society which do not have spatial form or which cannot be mapped? What does the answer suggest?

6. How is any agricultural system affected by spatial interaction? How and why might such effects differ among different types of agricultural systems?

7. How many different aspects of spatial form are conceivable, or mappable, for an agricultural system? a city? a particular culture? a manufacturing industry? a system of government?

8. Measure on an atlas map and state the straight-line distances as shown on the map between (*a*) Omaha and Des Moines, (*b*) Berlin and Baghdad, (*c*) Los Angeles and Tokyo. In the last case, compare this distance with that shown by measuring on a globe. A straight line on a globe whose plane intersects the center of the earth is called a *great circle* because if continued all the way around it would be a circle, as for example a complete meridian of longitude is. How do you account for the differences in the two distances as you have measured them? How is the difference related to the projection used in making the flat map which you consulted?

9. History and geography can both be said to deal with uniquenesses, that is, each historical period or each geographical area is to some degree unique, although it may have similarities with other periods or other areas. How does this affect the nature of historical and geographical analysis and the development of academic disciplines in both subjects?

10. Using an atlas map, find the large cities whose respective locations are indicated as follows, 40° north latitude and 75° west longitude; 4° 50′ north latitude and 74° west longitude; 34° S., 151° E.

11. State the latitude and longitude for the following: Berlin, Peking, Alma Ata, Denver, Honolulu, Manila, Capetown, Buenos Aires, Tonga Islands.

12. Why does the International Date Line not follow a straight north-and-south course along the line of 180°? What are the reasons for the deviations? What practical advantages are there for the location of the International Date Line in using Greenwich as the prime meridian?

13. How are latitude and longitude determined for any given position, at sea, for example?

14. New Zealand and Great Britain are roughly the same size, are both islands, and have similar physical environments. How can one explain their very different development, including the sizes and economic specializations of their respective populations, which come predominantly from the same racial and cultural stock?

Selected Samples for Further Reading (Chapters 2 and 3)

Abler, R., Adams, J. S., and Gould, P. *Spatial Organization: The Geographer's View of the World.* Englewood Cliffs, N. J., 1971. A good new introduction to the spatial analysis aspect of the discipline, including attention to the nature of science; a great variety of spatial patterns and techniques to study them.

Alexander, J. W. "The Basic-Nonbasic Concept of Urban Economic Functions," *Economic Geography*, XXX (1954), 248–61.

———. *Economic Geography.* Englewood Cliffs, N. J., 1963. A recent introductory text which combines some traditional treatment of spatial patterns in economic activity with quantitative and theoretical analysis and which also includes a simple treatment of central place theory, spatial interaction, and the analysis of urban patterns.

Bagrow, L. *History of Cartography.* Ed. R. A. Skelton. Cambridge, Mass., 1964. An authoritative and comprehensive survey of the development of map making.

Berry, B. J. L. "The Impact of Expanding Metropolitan Communities upon the Central Place Hierarchy," *Annals of the Association of American Geographers*, L (1960), 112–16. A case study and some general conclusions, of the spatial rearrangements in city functions which may result from improvements in intercity communications.

———, and Pred, A. *Central Place Studies: A Bibliography of Theory and Applications.* Philadelphia, 1961. Briefly annotated.

Brush, J. E., and Bracey, H. E. "Rural Service Centers in Southwestern Wisconsin and Southern England," *Geographical Review*, XLV (1955), 559–69. A useful application of central place theory to two different areas.

Buttimer, A. "Social Space in Interdisciplinary Perspective," *Geographical Review*, LIX (1969), 417–26. A thoughtful review of selected works on culture, imagery, and spatial patterns.

Chorley, R. J., and Haggett, P., eds. *Models in Geography.* London, 1967. A collection of papers by various authors, dealing largely with spatial patterns and interactions and their analysis.

Darby, H. C. "On the Relations of Geography and History," *Transactions of the Royal Geographical Society*, 1953, Publication No. 19. An excellent brief statement by a leading British historical geographer.

Febvre, L. *A Geographical Introduction to History.* Paris, 1925. A provocative and influential discussion of a close interrelation.

Forde, C. D. "Human Geography, History, and Sociology," *Scottish Geographical Magazine*, LV (1939). A useful discussion including a consideration of the holistic approach and the extent to which this is shared by sociology.

Greenwood, D. *Mapping.* Chicago, 1964. Popular, but accurate and clear, with a strong practical emphasis and a light touch.

Hagerstrand, T. *The Propagation of Innovation Waves.* Lund, 1952.

———. *The Diffusion of Innovations.* Trans. A. Pred. Chicago, 1968. The pioneer works in this field, with an essay by the translator.

Haggett, P. *Locational Analysis in Human Geography*. London, 1965. A good summary of the ideas and methods of the new generation of geographers concerned with logical and quantitative analysis of spatial patterns.

Johnson, J. H. *Urban Geography*. New York, 1967. An introduction to urban form, function, growth, and location, including cross-cultural material.

Morrill, R. L. *Migration and the Spread and Growth of Urban Settlement*. Lund, 1965. A clear and penetrating discussion of location theory, migration, and town growth, against the record of the past century in Sweden.

———. *Spatial Organization of Society*. Belmont, California, 1970. An excellent beginning text, clearly written.

Murphey, R. "Historical and Comparative Urbanization Studies," *Journal of Geography*, LXIV (1966), 212–19. A brief and simple outline of the major theoretical ideas relevant to the growth of cities, and a summary of the historical rise and spread of urban places in a variety of cultural settings.

Olsson, G. *Distance and Human Interaction*. Philadelphia, 1965. A comprehensive review article and bibliography on theories of location, spatial patterns and interaction, diffusion, and migration.

Pred, A. "The Intrametropolitan Location of American Manufacturing," *Annals of the Association of American Geographers*, LIV (1964), 165–80. A clearly written summary of the major patterns of industrial location within American cities, and the changes in these patterns over time.

Price, E. "Viterbo: Landscape of an Italian City," *Annals of the Association of American Geographers*, LIV (1964), 252–75.

———. "Central Courthouse Squares in the United States," loc. cit., LV (1965), 639 (abstract). This entry and the one above are studies of settlement form and their relation to culture, values, and functions.

Robinson, A. H. *Elements of Cartography*. 2nd ed. New York, 1960. Probably the best of the standard texts, clearly presented.

Sommer, R. *Personal Space: The Behavioral Basis of Design*. Englewood Cliffs, N. J., 1969. Urban planning and architecture and their uses—a discussion of a fascinating field.

Stewart, C. T. "The Size and Spacing of Cities," *Geographical Review*, XLVIII (1958), 222–45. A clearly written example of mathematical analysis in geography as applied to an important and interesting problem.

Thrower, N. J. W. *Maps and Man*. Englewood Cliffs, N. J., 1972. An excellent history and survey of cartography.

Tooley, R. V. *Maps and Map Makers*. London, 1949. A brief and well-written history of cartography and of the expansion of geographical knowledge.

Ullman, E. L. "Rivers as Regional Bonds," *Geographical Review*, XLI (1951), 210–25. The role of rivers and their valleys in shaping patterns of spatial interaction, with the particular examples of the Columbia and Snake rivers in the western United States.

———. "The Role of Transportation and the Bases for Interaction." In W. L. Thomas, ed., *Man's Role in Changing the Face of the Earth*. Chicago, 1956, pp. 862–80. Effects and causes of spatial interaction clearly outlined and applied to American examples.

Whebel, C. F. "Corridors: A Theory of Urban Systems," *Annals of the Association of American Geographers*, LIX (1969), 1–26. A persuasive alternative to classical central place theory as an explanation for the clustered pattern of urban places.

There is a great and increasing number of world atlases available, many of them of library size only. A few of the more practically sized and reasonably priced atlases are listed below. New and revised editions of most of them appear periodically.

Aldine University Atlas. A recent American compilation at reasonable cost, using high quality British cartography.

American Oxford Atlas. An excellent British atlas which sets a high standard.

Atlas of World History. Ed. R. R. Palmer. One of the more recent historical atlases, with good representative coverage of world history since the time of ancient Egypt and Mesopotamia.

Goode's World Atlas. Perhaps the best and most usable American atlas of comparable size and price.

Oxford Advanced Atlas. Another excellent British atlas.

Oxford Economic Atlas of the World. Primarily statistical, and extremely useful, with distribution maps of important commodities and resources and statistical tables for every country.

Oxford Home Atlas of the World. A smaller and much less expensive version of the *Oxford Advanced Atlas*, of easily portable size.

4
The Environmental Factor

THE SECOND FUNCTION of geography has been listed as a study of "the interrelationship between human society and the physical environment." The physical environment means to a geographer everything except man and his works: climate, soils, vegetation, wild animal life, landforms, the atmosphere, minerals, and water. The geographer calls these things the environment, or literally "the surround," because he is concerned with man. The word "environment" has meaning only in reference to some organic unit existing within it for which it provides the matrix or setting. It is obvious that the physical environment may have a profound effect on man and on how he organizes his activities. But it is appropriate to speak of an interrelationship because man in turn acts on his environment and alters both it and its effects on him. The environmental influence is thus far from being a constant, as man himself is not constant.

MAN REMAKES THE EARTH

Man changes his environment in a great variety of ways and has fundamentally remade the landscape of every area of the world which he has occupied in large numbers. In many cases succeeding periods of human occupance of the same area have each brought about a different change in the landscape, drained or irrigated it, deforested or planted it, enclosed it or left it as one great common. Man usually removes the forest first, although even this has not yet been accomplished in the taiga forests of northern Asia and North America or in most of the equatorial rainforest. Deforestation probably has more widespread effects on the rest of the environment and in turn on man than anything else man does. The agriculture which usually follows deforestation basically changes the soil or may seriously deplete it and again revolutionizes the appearance of the landscape. Depending on the nature of the agricultural system, woods and clearings are replaced by hedgerows and plowed fields, fences of varying kinds in each region, terraced rice fields, orchards, plantations, irrigated basins, vineyards, truck farms, pastures, open ranges, artificial lakes, dams, wells, ponds, or fields of grain. Man also builds a more or less dense network of roads and later railroads and airfields over the landscape, digs canals and dredges harbors, digs mines and oil wells, uses or exhausts mineral deposits, and reaps a harvest from the sea, changing it as well. All these changes in the environment react in turn upon man and influence his settlements and his activities. Because man has altered the environment and has learned to use it differently, he has thereby come under the influence of a variety of new factors—some positive, such as new resources, some negative or limiting, such as soil erosion—which were unimportant to him before or were important in different ways. Finally, his villages and towns grow into cities which in the modern period may cover many square miles with acre after acre of paved streets, factories, shops, and dwellings.

Where "nature" has been left, it has almost always been modified, at least in the environmentally favorable parts of the world where man has multiplied. Even the undestroyed temperate forests are not left alone but are systematically managed or replanted. Few parts of the modern earth outside the polar areas or the equatorial tropics would be recognizable to Neolithic or even to classical or medieval man. Animals kept or encouraged by men also play a part in modifying the environment. This is especially true of the grazing animals, and most of all of sheep and goats, whose close cropping prevents the regrowth of trees or bushes and even sometimes of grass. Over-grazing leads to serious erosion and may thus produce a change in the effective climate; many desert-like areas are in fact the work of goats. Man has been able to upset seriously the complex balance of interdependence among the biotic elements (animals, plants, birds, fish, insects) and the rest of the environment.

MAN MAKES HIMSELF

The force of environmental factors in their influence on man has led some investigators to give them undue weight, as suggested in Chapter 1. As usual with such deterministic approaches, there is a core of validity in the environmental approach and an idea of great usefulness, provided it is used intelligently. The physical environment is one of many factors which influence man. Political considerations such as subsidies, tariffs, central planning of the economy, or differences in the size of markets as between different political units may also be important or even predominant influences on agricultural systems, manufacturing patterns, or trade flows. Cultural factors such as taste preferences, traditional systems and techniques of agriculture or grazing, or systems of social organization may be more important than the physical environment in determining the type of land use or the nature of the economy. Just as importantly, man's technical abilities and wants are continually changing, and for all of these and other non-environmental reasons, man's use of earth varies enormously from place to place and from century to century.

Similar environments are used differently, different environments are used similarly, at different times and in different places. Similar cultures (see note 1 to the Introduction) may make similar use of different environments. Western Europe, southeastern Australia, and eastern North America tend to look alike as man has used and modified their landscapes, despite the differences in their physical environments, because they have been settled by people with similar cultures and similar techniques. By contrast, southeastern China and southeastern United States look very different, despite the close similarity of their physical environments, because of the great cultural and technical differences between their respective inhabitants, reflected in their use of the land. Chinese culture traits and associated patterns of land use have also spread into other parts of Asia and have created there what might be called a Chinese landscape. The Europeans, when they became active agents in the Far East, made a different use of the land and created a distinctive landscape in the form of plantations, as well as building cities which are European in appearance. Even in so rigorous an environment as the desert there are pronounced differences in culture and economy between arid areas which are physically very similar. For example, the deserts of central Asia, Australia, South Africa, and southwest United States all support different people with wide divergences in their culture and method of land use. Different people look for different things in the same environment and produce quite different results.

DETERMINANT OR INFLUENCE?

These obvious facts and the continuing changes in man's use of the environment in any one area invalidate the theory of environmental determinism. It is a tempting theory because it purports to explain the complexities and variations in human society in terms of a nonhuman constant. The physical environment may well be relatively constant, as opposed to human society, but its effects on man are far from constant and often far from clear or direct. Historians have remarked that the only thing which does not change is change itself. However, even this statement is not entirely true, since the process, nature, speed, and direction of change frequently differ widely, and to a large extent are reflections of the over-all growth or change of human society. Such an analysis applies also to changes in the environmental influence on man.

Nevertheless, that influence is profound. Few aspects of human society are free from it, although its effects may be disguised and are not often predominant. There is a risk that in discarding the theory of environmental determinism the baby may be thrown out with the bath water as happened to some extent when the unsoundness of environmental determinism was more clearly realized. Many scholars abandoned all attempts to consider the environmental influence or even to consider causality in general and confined themselves to simple description of spatial patterns and regional units without attempting to

was a serious loss; the
, or at least to ask why,
the environmental factor
y, which is to set man
nhabits; this necessarily
ations with his physical
which influence him and
unately, the discipline of
importance of that task.
eplaced by "possibilism,"
choice even though the
he can profitably do. Of
which environmental conditions
in a rdance not only with his
other influences or fac-
s political pressures, tra-
ment of particular tech-
to make one or a few
but choices may differ
at times. The rejection of
n of man's free will, but
ciety and its choices are
changing, that no one
omplex reality, and that
a concert of interacting
nportant as the physical
environmental influence
sometimes with striking
prominence, sometimes as a subdued note in the concert of influences in most aspects of human society and in its differences over time and over space.

In some ways, however, the use of the word *influence* is misleading; it may suggest, or even promote, an unconscious belief that the environment is an overt agent, that it acts upon human society. This, of course, is not strictly true, except in a very limited sense, as, for example, the destruction of property by floods or earthquakes. It is primarily man who acts and man who chooses. The physical environment merely exists. Although man may adjust to it or may follow certain courses of development in part because of the nature of environmental conditions, his actions are not in any direct sense influenced by the environment as if it were an active force. One speaks of environmental influences mainly because it is awkward and difficult to speak of the relationships in any other way without excessive use of words, but it is important to remember that the environment is passive and that man is the doer.

The nature of the environmental elements and their relation to man will be left for more detailed discussion in Chapters 7 through 13, where each can be considered separately. For the present it is enough to say that there is a fundamental relationship between man and nature and that geography attempts to delineate it, in its proper place, as part of the study of spatial patterns and the analysis of areal differences in human society.

CITIES AND THE ENVIRONMENT

Human settlements are not distributed over the earth in haphazard fashion but in a recognizable pattern, or in association with factors which may make a particular site or sites advantageous for settlement. Since cities exist in order to provide services for an area dependent on them, they flourish at locations which maximize access. If the landscape were a flat, undifferentiated plain without rivers or local differences of any kind, distance would be the only factor and the city would be located in the exact center of the area which it served. But landscapes are never entirely uniform. There are local differences in topography, soils, ground water, economic productivity, culture, political pressures, and many other factors which tend to make certain sites advantageous and which may in particular make access easier. Cities frequently grow up near points where two different physical areas come together, such as mountains and plains, land and water, forest and prairie, or desert and humid areas. Study of an atlas will show how frequent this association is, and an example of it has already been examined in the case of Timbuktu. Cities are centers of exchange, and exchange implies difference. Where differences meet is a convenient place for exchange; transportation is minimized in both directions. In moving from one kind of physical area to another, it may also be necessary to change the mode of transport, as at Timbuktu.

Cities of the Fall Line

From Providence, Rhode Island, through Montgomery, Alabama, lies a string of cities arranged in a line roughly parallel to the coast but in most cases removed from it and at the inland edge of the coastal plain where the plain merges with the foothills of the Appalachian Mountains. Along this line the physical conditions of the coastal plain change and are replaced by those of the uplands or the Piedmont, as it is called (literally, "foot of the mountains"). Among the cities which punctuate the line are some of the largest in the world, and most of the others are of more than local importance. The list includes Providence, Fall River, Hartford, New York, Trenton, Philadelphia, Baltimore, Washington, Richmond, Raleigh, Columbia, Augusta, Macon, and Montgomery. Why are they there? What are the advantages of this zone for access and exchange, the basis of nearly all cities?

A median position between two different physical regions involves many advantages. But there is also a more specifically localizing factor which is related directly to the local environment. The piedmont is composed of relatively hard rock, the coastal plain is relatively soft or sandy, and the rivers flowing from the mountains to the sea have cut into the coastal plain along this line more deeply than into the hard piedmont rocks. Nearly every river along the line is broken by falls or rapids as it tumbles down onto the plain, or at least navigability sharply decreases, and above this point navigability is much less than below it. Below the falls many of these rivers, especially north of Richmond where the largest cities are, widen out markedly because they have in effect been "drowned" by an invasion of the sea and are really estuaries or bays rather than rivers, deep and wide enough for ocean-going ships.

The "fall line," as it is called, offers four specific advantages as an urban site: (1) local water power; (2) the lowest point at which the river can easily be bridged or crossed and where land transport routes are thus bunched together; (3) the point at which goods arriving from the sea must be unloaded onto a different carrier, either by land transport or a smaller river boat; and (4) a place of vantage on a river which gives access to a tributary hinterland lying in and behind the mountains through which the river's valley cuts.[1]

The first two advantages are particularly important in the early establishment and growth of a city even though they may lose their importance later on. London, for instance, grew up as the lowest practicable bridge site on the Thames, and it remains so. Fall River, as the name implies, based its early industrial leadership on local water power, as did Trenton. But it is the last two of the four advantages which give a sustaining basis to a city, because as economic development proceeds, they increasingly emphasize the city as a focus of exchange. Wherever carriers must change, the process of exchange is most easily concentrated. Goods must be unloaded from one carrier and loaded onto another. They may be stored before moving on, or they may be processed or manufactured. In any case, such a point is a convenient collection and distribution center. These break-in-bulk points, as already discussed, are admirably illustrated by the cities of the fall line. The break comes there rather than on the coast because water transport is cheap and is therefore used as far inland as possible, because location in the zone between plains and hills makes the fall line city a more convenient exchange center, and because the river itself, or its valley, provides an easy route into the interior.

Usually one city in a region has the best combination of advantages and becomes much the largest. New York is a convincing example; although the fall line or the head of ocean navigation is farther up the Hudson, New York's magnificent natural harbor on the sea itself is a major asset. But the advantages of the fall line are enough to support many other cities, too, including Philadelphia, only about ninety miles from New York but still among the ten or eleven largest cities in the world. Like New York, Philadelphia is not supported directly by the fall line. On the Delaware River the sharpest physical break comes farther upstream at Trenton, in the form of rapids, although Philadelphia was the head of ocean navigation until the recent dredging of the Delaware above the city enabled large ore-boats to reach the new steel plant at Morrisville, just below Trenton. The Schuylkill, a smaller river flowing into the Delaware at Philadelphia, has its falls within the city, however, not far from the central business district, and the Schuylkill valley provides an easy route to the west, including important coal-pro-

---

[1]The head of ocean navigation and the geologic fall line may be fairly far apart, and neither may be marked by an actual waterfall.

ducing areas. Philadelphia is supported by the market and productivity of a large inland area which the advantages of the fall line help to make tributary to the city. Actual distance is, of course, an important part of access. New York had the advantage of being more or less in the middle of the most populous and productive part of the coastal plain. Philadelphia was close to the middle, but Baltimore, Washington, or Providence were farther from the main population center and suffered accordingly in the competition.

Physical conditions often closely affect the works of man. On the eastern seaboard human activities, including railroads and highways as well as cities, have focused on key points along the line between plain and hills where carriers must change and where there is maximum access to the largest area.

The Fortunes of Washington

Washington, D.C., was selected as the site for the national capital in part because of George Washington's advice, based on his perceptive appreciation of the site's advantages (see Map 6). Not only did it lie close to the fall line (the falls of the Potomac, an impressive sight, are close to the outskirts of the modern city) and at the head of the large estuary of the Potomac River, but it commanded what was then the easiest and most traveled route from the major area of colonial settlement on the coastal plain to the new areas being settled and developed across the Appalachians. George Washington saw that a city near the mouth of the Potomac might make use of the natural route through the mountains which the river had carved and along which the famous Cumberland Road was built, perhaps to become the greatest metropolis in the country. To support his idea he encouraged the building of a canal (which can still be seen) following the river through the mountains, with its terminus at Cumberland, Maryland, which was to funnel the trade of the west into Washington.

A generation after Washington's death began the revolutionary development of the railroad, which he could not have foreseen but which quickly made his canal obsolete. Baltimore, an old established trading city with capital to spend and ambitions to further, built the first major railroad in the United States, the Baltimore and Ohio, along the Cumberland route and the Potomac valley, which effectively short-circuited Washington. Hard on its heels came the Pennsylvania Railroad from Philadelphia, an even larger city, which was not going to be surpassed by Baltimore. Although the route chosen was a difficult one, due west from Philadelphia across the successive ridges of the Appalachians, this was not a major problem for a railroad. The Pennsylvania became the largest railroad of all, and indeed the largest in the world, in the volume of its traffic, because its eastern terminus (quickly extended to New York) enabled it to serve the largest number of people, and because coal and oil were discovered and produced along the mountain route of the Pennsylvania, providing it with a large volume of traffic flowing also from the major industrial operations which arose in association with the coal. A little before this, settlers in New York had finally managed to clear the Mohawk valley of the hos-

MAP 6

WASHINGTON, D.C. AT FALL LINE AND AS PASSAGE POINT TO APPALACHICAN PLAIN

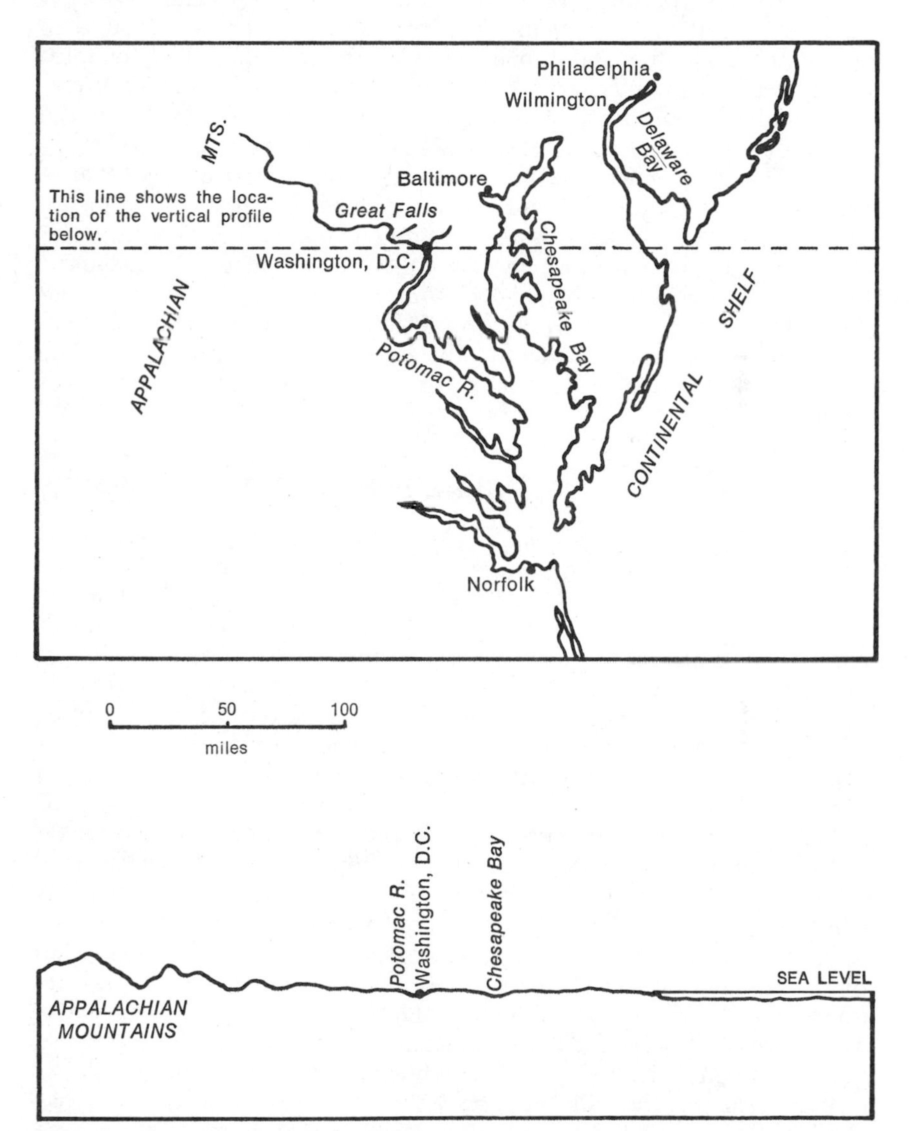

The scale above applies only to the horizontal distances in this profile. The vertical distances are exaggerated approximately twenty times.

tile Iroquois Indians who had blocked it in Washington's time. The Erie Canal through this lowland gap, using a much easier and more productive route than Washington's canal, was followed by the New York Central Railroad, soon a close second to the Pennsylvania in the volume of its traffic. In both cases the volume of flow was and is a reflection of the productivity of the country through which the routes pass, as well as of the termini. The productivity was in part due to environmental factors and to man's new ability to make use of the middlewestern environment and its resources. But it was New York and Philadelphia rather than Washington which profited most from the growing productivity of the Middle West because their access to it was better. The environment meant something different to a railroad than it meant to a canal; railroads could be built across mountains. The meaning of the environment is always changing as man's needs and abilities change. Railroads reduced what is called "the friction of distance." Railroads, political and financial factors, the increasing size of ocean-going ships, and other changes largely ruled out the particular commercial advantages which George Washington saw for his capital. The environmental factor was by no means eliminated, but its influences changed.

THE ENVIRONMENTAL INFLUENCE

It is when one selects a single element of the environment, such as climate, and attempts to base complete explanations on it that the risk of overemphasis is greatest. Climate may be largely responsible for the Eskimo's way of life and for the lack of population in the desert. It is a temptation to use the same simple key to unlock the more complicated puzzle of man's use of the land in more temperate parts of the earth. It would be strange if such an obviously limiting or conditioning factor as climate, one which varies so greatly from place to place, did not have a discernible part in influencing the pattern of human society. But this does not mean that people who live under certain climatic conditions are bound to think or to live in a certain fixed way, or that the climate of some parts of the world gives to their inhabitants a distinctly "progressive" or "backward" character, as many climatic propagandists would have us believe. The environment usually operates as a whole. To isolate one of its factors and ignore the others is to do violence to the actual situation, where there are a host of nonenvironmental factors operating as well, and where technological change is also taking place. Man's own efforts are fully as important as the physical characteristics of the earth on which he erects his civilization. A man standing on a hill is there because the hill supports him, but equally because he has climbed there himself.[2] The physical environment is important as one of many influences on man acting in concert, not as a determinant. It is both difficult and misleading to examine the environmental influence without continual attention to other influences.

---

[2]This illustration was often used by the prominent British geographer Sir Halford Mackinder in attempting to explain the dual nature of geographical analysis.

NATURE AND MAN IN SWITZERLAND

The environmental factor is stressed in the following example in order to make a clear illustration, but the example would not be understandable unless spatial relations and a variety of nonenvironmental factors were also considered.

Switzerland is for the most part a highly mountainous country close to the populous lowland area of western Europe. Most of the Swiss population, the cities, and the manufacturing are concentrated on a central plateau, but this area with its famous lakes is surrounded on all sides by high mountains, and there is very little level land even on the plateau. The important basic goods and services produced by the Swiss economy are listed below, and a list of Swiss cultural characteristics follows:

Goods

dairy products and cheese
chocolate
watches and clocks
wood carving
optical goods
lace and embroidery
machine tools
fine textiles
precision machinery
electric generators
electric furnace steel
chemical and pharmaceutical goods

Services

boarding schools
tourism
sanitaria
banking

Cultural Characteristics

political neutrality
four languages, but no Swiss language
ancient nationalism
long-standing democracy

Goods

There is a quality common to all of the goods listed: high value per unit of weight and a generally high standard of excellence. Beginning at the top of the list, dairying is frequently practiced in areas which are marginal for field agriculture, usually because of slope, unsuitable or stony soils, or cool, cloudy climate. Such areas may, however, support adequate or even good pasture, and the slope is only a minor disadvantage for animals. Dairying is prominent in Wisconsin, Scotland, New England, and New Zealand in part for these reasons. The Swiss mountains offer a more profitable basis for dairying than for field agriculture, and in fact in the high Alps (a word whose original and proper meaning is a high mountain meadow) there are fine pastures. They are free from snow only in summer, hence the general Swiss practice of *transhumance,* or the seasonal movement of animals and people up to the Alps for the summer grazing, from which they come down to lower levels for winter residence. In winter the animals are fed on barn-stored hay, cut the previous summer in the Alps. The Swiss use goats as well as cattle, since goats are better able to subsist on the steep slopes and less luxuriant pasturage, but they have also developed a type of dairy cow, the Brown Swiss, whose exceptional hardiness fits it to its environment without unduly sacrificing the quality of the milk.

The Swiss have an obvious problem of transportation to get the milk to market, however. Switzerland itself is not densely populated, and its major markets lie outside the country in the European lowland. Fluid milk is bulky in relation to its value and presents special and expensive transportation problems. The Swiss make cheese, converting the milk into a concentrated and high unit-value product which can bear the cost of transportation across mountains. This solution is commonly adopted by dairying areas remote from their markets. Closer in, butter may supplement cheese or take its place, but fluid milk is dominant only within easy reach of the market. This areal pattern is readily observable in Wisconsin in terms of the distance from the urban area of Chicago-Milwaukee, or of Minneapolis-St. Paul. Switzerland's linear distances to its market are comparable, but they are more expensive distances because of the mountains. The cheese made in Switzerland is of unusually high value per unit of weight; it is very concentrated, as in the case of "Swiss" cheese (the kind with holes in it), or of fancy high quality, as with Gruyère, each piece individually wrapped in foil and packed in attractive little boxes. Switzerland has no particular advantage for dairying or cheese-making in comparison with more favored lowland dairying areas nearer to their markets in terms of cost; it must compete on the basis of high quality and the reputation which it wins.

Chocolate-making may also be related to the mountains. There are no cacao trees in Switzerland, and the raw chocolate must be imported from the equatorial tropics where it grows. But there is plenty of milk, usually more than can profitably be made into cheese since milk quickly spoils. The Swiss make milk chocolate and convert what otherwise might go to waste into a very high unit-value product which has established a world-wide reputation for quality. Milk chocolate can be made more cheaply in Hershey, Pennsylvania. The Swiss transportation bill for inbound raw chocolate and sugar and outbound finished product is much higher; their product must be correspondingly better in the sense that it can bring a higher price.

Swiss watches, perhaps the most famous items of all, emphasize dramatically the solution which this mountain area applies to its transportation problem. During part of the Second World War when Switzerland was a neutral island in Nazi-dominated Europe, it was allowed to make one monthly shipment to the outside by road as its export quota. The dilemma of what to send was easily solved by filling an automobile with Swiss watches for export; the monetary value of such a shipment strains the imagination, but the cost of haulage was, of course, negligible. Watchmaking also gets around the problems induced by Switzerland's virtual lack of mineral resources. The value of the raw materials used in watches is a tiny fraction of the total finished value, most of which has been added by manufacturing. It pays the Swiss to import the highest quality Swedish steel; they can sell hairsprings made from this steel, for example, for well over a hundred times what it cost for the steel, by weight. The best steel may cost $5.00 a pound; Swiss hairsprings or watch movements may sell for $50.00 an ounce.

Long experience and skilled labor help to strengthen the Swiss position of leadership in this industry. As their advertisements put it, "Time is the art of the Swiss." Winter is a slack season in a land with heavy snowfall and rigorous temperatures. Relatively little labor is needed to pitch stored hay down to the animals in the barn or to manage the other wintertime aspects of the dairying economy. Something was needed to fill this slack season, profitably if possible. The women made lace and embroidery, and the men did wood carving, as the Nantucket whalers used to carve whalebone and ivory between bursts of activity on their tediously long sea voyages. Little figures, houses, and clocks with wooden works were among the Swiss woodcarving specialties, including the still-famous cuckoo clocks and other precision gadgets, made originally as amusing toys and ultimately for sale. From this kind of beginning came the tradition and skill which go into Swiss watches.

Swiss concentration on high unit-value products and on value added by manufacture is clearly carried out in the cases of optical goods, machine tools, fine textiles, precision machinery, pharmaceuticals, and electric furnace steel, all produced in the cities or along the fringe of the central plateau. In the making of generators use is being made of Switzerland's advantage, as a well-watered mountain area, for the development of hydroelectric power, but production is also concentrated on high-quality goods. Electric furnace steel is expensive but is of high quality, and in Switzerland it helps to supply the precision metal industries, including watchmaking. Imports of bulky raw materials are minimized, and the emphasis is placed on the value added by manufacture, based on skill, tradition, care, and hard work. Most of the goods which Switzerland produces are related to the fact that Switzerland is a mountainous country near the center of the European market.

Services

In providing services, the Swiss have translated the handicap of the mountains into a profitable asset. Tourism as a major enterprise is relatively young, not much more than a century old on any significant commercial scale. (The extensive religious pilgrimages of medieval

Europe and the Islamic world are in rather a different class.) The Swiss, if not the inventors of tourism, were among its earliest modern pioneers. Tourists provided a perfect answer to the transportation problem, especially as average incomes rose in Europe after the eighteenth century and transport facilities improved so that Switzerland was easily accessible and could begin to let tourists pay the cost of developing its transportation network. Tourism became Switzerland's largest single source of income, and the Swiss established a wide reputation as hotelkeepers, makers and demonstrators of sports equipment, guides, and suppliers of fashionable vacations. Other mountain areas have since found such specialization profitable, including Japan, Norway, and northern New England. School pupils and medical patients could be attracted to Switzerland for similar reasons. In education and medicine the Swiss also established a high reputation for performing essential functions with particular skill against the therapeutic or inspiring background of the Alps, whose value finds its way into the school and sanitaria bills.

Environment and Spatial Interaction

The force of the environmental influence in Switzerland is clear, but Switzerland's central location within Europe was also an essential condition for the development which has occurred. Tibet is doubtless more impressive, but few people consider spending their weekends or even summers there. If they went, they would not find the highly skilled specialization of the Swiss in services like medicine, tourism, or hotel-keeping, which are themselves reflections of Switzerland's accessibility both to European customers and to the ideas and techniques which are easily spread from all quarters of lowland Europe to crossroads Switzerland. Though mountainous, Switzerland lies athwart important trade routes and is surrounded by several separate major European cultures whose enriching influences have contributed fundamentally to Switzerland's development. The growth of a vigorous, sophisticated, productive culture in any area may often be related to the degree of access and of cross-fertilization or external stimulation which the area enjoys. Switzerland is virtually unique among mountain areas in this respect. The Alps are pierced by several relatively easy passes, and although heavy goods are still at a disadvantage, trade and ideas move freely in large volume and variety. Switzerland separates three major and complementary physical and cultural regions: the Mediterranean Basin, northwest Europe, and central Europe, which exchange with each other to a large extent through Switzerland. Switzerland is what French geographers call a *terre de passage* (land of passage).

The political independence of modern Switzerland and its prosperous economic development have been said to date from the thirteenth century. At that time a new and remarkable stone bridge was built across the gorge of the Reuss River (known as the "Devil's Bridge," since tradition held that the Devil's aid was necessary in building it) and a suspended causeway constructed along the sheer sides of the gorge, which greatly increased the usefulness of the St.

Gotthard Pass, then the easiest of all the Swiss trans-Alpine routes. Certainly the flow of trade contributed basically to the development of the Swiss economy and helped the country to maintain its independence. But considerable trade existed before the approaches to the St. Gotthard Press were improved, and it was only for this reason that it paid to make the improvements, just as it paid to dig several railway tunnels under the Alps with the greatly increased trade of seven centuries later. The Po plain of north Italy and the Rhine region of north Europe, separated by Switzerland, are environmentally and culturally different trade-generating centers of a complementary nature; trade flows through Switzerland despite physical obstacles. When the obstacles were lessened, of course, trade flows increased even more.

A parallel may be drawn here with the routes across the Appalachians. The Pennsylvania Railroad "defies" the mountains because they separate productive complementary areas whose trade easily paid for the building of the railroad, as trade volume had earlier paid for the Cumberland Road; the areas now joined by the railroad can exchange in greater volume and variety because of the improvement in transport. Many physically easier routes elsewhere are not used, or are used much less, because they do not join productive complementary areas. Similarly, it was not simply the Erie Canal or the New York Central Railroad which made New York City grow or which produced the trade which feeds the city. The trade flowing between the Middle West and the east coast led to the building of the canal and the railroad as well as to the growth of the city. Switzerland as a whole may also be compared with Britain, where continual cross-fertilization from varied continental cultures helped to produce a vigorous and dynamic British society which also depended on world-wide trade routes leading to the English Channel. Holland's spatial relations were similar: external stimuli were strong, and a national state arose without the protection of physical barriers but nourished by extensive and profitable trade at the mouth of the Rhine. In Switzerland, trade resulting in part from Switzerland's location helped to build the state and the economy. Mountains made it possible for an independent state and economy to be maintained, as well as giving Swiss culture and economy a particular character.

Neutrality

The service of banking is hard to relate directly to the influence of the Alps. Switzerland is relatively central within Europe, but the success of banking depends more on Switzerland's political neutrality and on the flow of trade through the country. Neutrality made Switzerland an essential clearing house among the belligerent European powers and gave a greater degree of security to funds held in the country. Switzerland is poor in domestic resources and heavily dependent on foreign trade to make up its deficiencies and maintain its people; trade flows resulting also from complementary exchange between north and south Europe provided a further basis for banking and the financing of trade. But neutrality was an important asset. How can it be related to the nature of the Swiss environment?

Neutrality is hard to come by in Europe. Most of the smaller nations would prefer not to be involved in Europe's wars, knowing that they have little to gain and much to lose. But most of them have been unable to resist the military pressures of the Great Powers, and after a futile defense of their neutrality have succumbed in nearly every modern European conflict. Norway, Holland, Denmark, Belgium have all been thus overrun. Only Switzerland (and to a lesser extent Sweden) has been able to hold aloof. A neutral clearing house is useful to belligerents, and Switzerland is conveniently near the middle of Europe. But more importantly, the Alps offer few advantages to an occupying power. Norway, Denmark, Holland, and Belgium all gave to a continental belligerent a foothold on the coast from which the sea lanes could be menaced or dominated or an attack on the continent resisted. Some of the small countries also had heavy industrial establishments of direct military value. In mountainous Switzerland there was nothing, or at least nothing of any use to a belligerent adequate to recompense him for the effort of conquering a country whose natural defenses were so formidable. The Swiss have always been prepared to defend their country. The highly efficient Swiss militia has led to the saying that "Switzerland has no army; Switzerland is an army." With the help of the Alps, Switzerland would be a hard nut to crack. But any major continental power could have cracked it, if the gain had been worth the effort. Since it was not, Switzerland was left alone and used its neutrality profitably. In addition to banking, the country also acted as refuge for dissident or persecuted groups from all of Europe, most of whom were economically valuable assets, and it has more recently become the headquarters for a variety of international organizations. Swiss neutrality is carefully preserved. Switzerland declined to join the NATO group for this reason and has even remained outside the United Nations. The Swiss feel they cannot afford to alienate any power by allying themselves with another, whatever their own political sympathies.

Nationalism

Nationalism, in the sense of regional distinctiveness and communal identity, is old in Switzerland, at least as old as the time of Julius Caesar, who speaks recognizably of the Swiss as a national group under the name of Helvetians (still used in Switzerland) in his *Gallic Wars*. Most of the present national units of Europe were unformed at that time and have gone through a series of changes since. Mountain cultures are not easily affected by technological and social change, and they occupy an environment fundamentally distinct from the lowland areas around them. The people who lived in the mountains felt a common bond and strove to distinguish themselves from non-mountain people. They earned their living in a "mountain" way and for other apparent reasons were a people apart. Love is seldom lost between mountain and lowland peoples, and where such areas are juxtaposed, pronounced human differences may result, often expressed in the form of political separation or passionate regionalism. Modern Switzerland began as an independent nation-state in the form of three mountain-forest communities called "cantons," which gradually increased their autonomy within the Austrian empire to the point where they could by force of arms complete the separation from Austrian control. The mountains have given the Swiss an advantage in fighting for their independence or neutrality, and they have long been famous military practitioners, famous enough so that their services were for centuries in demand among all the great powers as mercenary troops. A poor area, especially a mountainous one, often sends out a stream of emigrants seeking better opportunity elsewhere; Swiss emigrants had their military prowess to offer for sale as their most marketable service. Comparison with Scotland and Nepal (home of the Ghurkas) suggests that this is a common mountain practice.

Language

It is in the light of Swiss nationalism and community of interest, related to its mountain environment, that one can best interpret the country's anomalous linguistic situation. There are other bilingual or multilingual states, but with them this difference is a source of friction, as it is for instance in India or Belgium. Nowhere has a modern national state approaching Switzerland's small size existed with no language of its own and with four languages in common use. The area of contemporary Switzerland has been successively detached from the surrounding states: Austria, Italy, France, and what is now Germany; hence the origins of its languages. The place names on any map of Switzerland give some evidence of linguistic areas. The mountain community of interest was so strong that not only did these linguistic differences offer little obstacle to unity, but there was no need for a single common language to hold the community together. Italian-speaking residents of the Ticino district are as fervently Swiss as the German-speaking residents of the three original cantons, Uri, Schwyz, and Unterwalden, or as the French-speaking people of the Geneva-Lausanne area.

Romansch is an archaic language which has disappeared everywhere else and survives only in a small part of Switzerland. The mountains have sheltered Romansch, given it an area of refuge where it did not rub up against competing influences, and it has been preserved also in part as a badge of regional distinctiveness, much as Welsh and Gaelic have been in their mountain settings. Mountains are nearly always refuge areas for relict cultures or cultural traits, and despite Switzerland's intimate economic, cultural, and political involvement with the rest of Europe, probably on a greater scale than any other mountain area, the remote Romansch-speaking corner of Switzerland remains a backwater in at least this one respect.

Social and Political Structure

Democracy, in the sense of individualism within a cooperative framework, is understandable as a feature of a hardy pioneering society. In Switzerland, as in many other mountain settings, Scotland, for example, individualism is often expressed in a pronounced clannishness and is reflected in the federal system of government in Switzerland despite its tiny area and relative homogeneity. Common interest may often best be served thus, but individual self-reliance is also an important element. However much the mountains may have accentuated such traits, mountains were also involved in a less direct way in two respects of perhaps greater importance. A major basis of the old stratified feudal society of Europe has been ownership of land. There was little good agricultural land in mountainous Switzerland, and there was never a significant split between the aristocracy and the lower orders in Swiss society. Actually, part of the Swiss movement for independence centered on objection to the stratified and aristocrat-dominated society of medieval Austria; nobility and royalty have ever since been banned from Swiss society, except, of course, as foreign tourists with money to spend. Later, as the modern industrial revolution progressed, heavy manufacturing could not economically establish itself in the Alps, and change was concentrated in the precision industries, where skill in the artisan tradition was more important than mass labor. There is still a relatively small spread in incomes among the Swiss population as a whole, and there is little social distinction on economic or occupational grounds. The division between industrial proletariat and industrial capitalists may not prevent the growth of a healthy democracy, but its absence certainly helped to preserve the original egalitarian and democratic nature of the Swiss community.

The Alps make a dramatic environment, and one might expect that the effects on the people who live there are both larger and more readily discernible than in the environmentally more moderate parts of the world, which after all are where most people live. Switzerland is, one may say, halfway between a situation like the North Pole, where man has little choice, and the temperate lowlands, where the environment presents him with a wide range of choice. But the force of the point made by the example of Switzerland is nevertheless applicable to all of man's relations with his physical world. Although the interrelationship is not often so clear, there are very few situations or developments where a causal connection between man and nature cannot be traced.

5
The Ecological Imperative

MAN THE DESTROYER

MAN HAS BEEN fighting against his environment ever since he came into existence. What distinguishes man from the other animals is in fact primarily his ability to manipulate the environment, first (perhaps half a million years ago) by the use of fire, then by progressively more effective hand and mechanical tools, and finally by a great variety of machines and chemicals. The last hundred years have seen this process accelerate tremendously, as man rapidly increased his abilities to manipulate the environment through industrialization. He has probably made more destructive impact on his environment during the past century than during the whole preceding period of human history, an impact which is still continuing in full force. The change is a revolutionary one, and one which has only recently and belatedly been recognized as posing a serious threat to human survival itself.

In the simplest terms, ecology is the study of the mutual interplay among and between organisms and their environment; originally a biological concept only, it has recently been expanded to include, and now to emphasize, the role of the human organism and man's interrelationship with those parts of the natural environment which are important to him. In this context, the nature of the environment as a set of interlocking systems in balance with one another is especially clear, a balance easily disrupted by the aggressive powers of man, the part of the ecological system which has grown out of control.

The natural environment is a fragile construct, or at least it is dangerously vulnerable to the destructive manipulation which man has practiced on it. Other species have adapted to their environment or perished. Man has survived a major environmental change in the form of the last glaciation with the help of his new weapons such as fire, but his newer weapons have come to threaten the very base which sustains him. On a human time scale, the environment changes very slowly as a result of its own processes. Man has made himself in effect an external force acting on it, in such a way as to produce rapid and sweeping changes. He has divorced himself from his environment and as a result may have cut the cord which nourished him. Without this new external pressure, the environment would in time heal itself, but the pressure of modern human technology is unrelenting and still increasing. It continues to disrupt the comparatively slower and more fragile processes by which the various elements of the environment remain in a dynamic balance.

Awareness of the dangers involved in disrupting the natural processes is very recent, except for a few far-sighted individuals who, in the early stages of the industrial revolution, realized some of its destructive consequences and warned against them. Like most prescient people in any age, they were ignored. Their warnings were easily put down in the atmosphere of enormous confidence in man's supposedly unlimited powers brought, understandably, by industrialization. The Victorian and post-Victorian West were supremely confident that Western man had created a new kind of "progress," and through his new industrial technology of steam and steel had finally "conquered" nature. That confidence has begun to be undermined only in the last few years, and it takes a great deal of time to slow down, let alone reverse, almost any aspect of mass behavior, especially one which has been supported with the enthusiasm given to industrial technology and its fruits. Nor can it be denied that industrial technology has, at least in the short run, tended to improve the material basis of life for most people in the areas where it has spread, including its effects on public health and food production.

The longer-run effects of the internal combustion engine, the use of chemicals in agriculture, food processing, and industrial products, nuclear weaponry or testing, and the burning of fossil fuels (coal, oil, gas) may, however, tip the balance sharply to the negative or even the fatal side. Awareness of these risks is still so recent that we really do not yet know their dimensions, nor how best to guard against them, short of abandoning the whole structure of industrial technology. Such a solution is of course unrealistic even if it were appropriate, but each year it becomes clearer that a drastic change of some sort in human behavior must take place, and as rapidly as possible, if man is to avoid altering his environment to the point where he can no longer survive in it. He cannot of course eliminate it; he can, however, eliminate, or disrupt or destroy those parts of it which sustain him.

ATTITUDES AND FASHION

Public concern over this problem, however belated, is encouraging. For it is only when large numbers of people become excited about an issue that mass behavior and public policy are likely to change. But the current ecological uproar does unfortunately have an element of voguishness about it. It became the fashion among a few in-groups in the late 1960's, and by the early 1970's among large sectors of the general public. Much of this uproar has not been accompanied by serious thought or study, and much of it has merely repeated catch phrases, jargon, or symbolic but not very significant changes in behavior, such as the craze for low-phosphate washing agents. Unthinking commitment to what appear to be the most dangerous aspects of industrial technology has been affected to a minor degree. These things include the unbridled use of the private car, nuclear explosions, defoliants, plastic containers, poison sprays, and food additives. Most people are not yet at the point where they are willing to give up any significant personal convenience, or established habit patterns, in order to slow down what may prove to be disastrous or even irreversible destruction of the environmental base for future human life. The ecological vogue may wane as rapidly as it appeared, like so many other vogues; most people are notoriously short-sighted.

It would be comforting to be able to look to the academic disciplines for reassurance and continuity, but they are often the most vogue-ridden groups of all. Ten years ago or less one heard or read very little about environmental deterioration except for a few voices in geography or natural resources and a few writers for the general public whose books were often dismissed as scare stories. The word *ecology*, which connotes a perfectly valid concept of the organic interrelationship among living things sharing the same natural environment, was rarely heard; in the past few years it has become a rallying cry in many academic disciplines as well as in the popular press. It is often misused in both places, or used with such vague generality as to be almost meaningless.

It is reassuring, however, that a number of academic disciplines have turned their attention increasingly to the vital interrelationship between man and his environment, and to the potentially or actually disastrous consequences of many aspects of human behavior in this connection. Anthropologists in particular have been concerned for some time with the ecological nexus and with nature-culture interactions in their regional variety. This concern has been taken up by more and more anthropologists in the past few years and seems likely to present an enduring (if overdue) change in the discipline. Historians, sociologists, political scientists, and even a few economists have also picked up this theme, and one may hope that they will persist with it.

The natural sciences have always been concerned of necessity with ecological analysis, and indeed their work has been primarily responsible for the existing body of knowledge, but in recent years they too have given increasingly broader attention to the outstanding human dimension of the problem. Geographers cannot afford to look down their noses at others, however, since although in principle they have always had a commitment to environmental study, the overwhelming bulk of geographical work between about 1940 and the end of the 1960's represented anything but such a commitment; studies of the man-environment interconnection are still only a fraction of all geographical work, although they are rapidly increasing. Some of this reluctance to address what would seem to be a central concern of the discipline doubtless stems from the earlier reaction against the extremes of environmental determinism; but more than a generation has passed and it is now no longer out of fashion for geographers to consider, as the nature of their discipline requires, the interrelation between man and his natural world. In an age of galloping industrial technology and headlong global population growth, there can be no more important problem for study.

MAN'S CONQUESTS

For most of his time on earth, man has regarded nature as an enemy, or at best as something to be either feared or overcome. Man began as a relatively defenseless creature, surviving more by his wits than by his physical abilities. It is understandable that he saw the ecological nexus more as a contest than as a secure or mutually adapting partnership. As his inventive powers increased with cumulative experience, nature became more and more a tool as he was able to make more of it serve his needs. It could also be viewed as the handiwork of man's religious construct, an all-powerful Creator, who in many religious traditions was seen as having created the earth in order that it might serve God's supreme creation, man himself. Folk religions might see the divine also manifested in natural phenomena, and might even worship a mountain, a stream, or a forest. But few cultures did not exalt man in their own cosmos to a status equivalent to the lord of creation, and few did not in practice, whatever their philosophical positions, attempt with varying degrees of success to alter their own environment, often fundamentally, in order to make it serve man better and with little regard for either short or longer run ecological balances. As discussed in more detail later in this chapter, even the Chinese did not allow their views of man's relative insignificance in the natural universe or of the importance of harmonious adjustment to nature prevent them from almost completely deforesting their country, killing off most of the original fauna and flora, and transforming the landscape into probably the largest artificial agricultural system in the history of the world.

With the advent of industrial technology on a large scale in the latter half of the eighteenth century in the West, views of nature as tool or obstacle became still more prominent. The "conquest of nature" became a familiar phrase, and most people believed that it was finally being accomplished. Man's new powers were indeed awesome, and they were continuing to grow more rapidly all the time. Confidence was further increased by the spreading support for the ideas of Charles Darwin, or more accurately those of some of his followers and contemporaries, such as Herbert Spencer, who applied Darwin's theories about the survival of the fittest species to the struggles of "civilized" man against the forces of nature. Such thinkers and their supporters in fact misapplied Darwinian theory, which dealt with an enormously greater time span of evolutionary development and with species other than man, but there were many proponents of "social Darwinism," even until very recent years, who urged in effect that ruthlessness was essential to survival and that only an aggressive and conquering human species could maintain itself in what they often saw as a hostile non-human, or even inter-human, world.

As industrialization entered the scene, however, there was an accelerated shift in resource use and in sources of energy, from organic to inorganic or fossil (such as coal and oil), and a revolutionary increase in the scale of man's impact on his environment. Previous resource use had centered more importantly on agriculture, which in a sense nourishes the earth to make it more fruitful, and on a much more limited use of fuels and energy sources for household purposes or for small-scale manufacturing. Mineral resources were relatively little used, and relatively large areas of the earth were thinly inhabited and very lightly drawn on in a resource sense. Industrialization relatively suddenly brought a conglomeration of overwhelming new pressures on the environment, most of them with seriously disruptive or destructive effects, and many of them involving previously under-used areas. Industrialization has for example profoundly altered the global distribution of population and economic activity, as well as grossly multiplying population totals. These problems involving the shift in resource use patterns and the interrelation with population growth are discussed in greater detail in Chapter 13. But precisely because man's technical powers over the environment had increased so dramatically, because he could now operate steam-driven machinery to dig mines, clear forests, build roads and railways, transport raw materials great distances, manufacture and use huge amounts of chemicals, and support a rapidly growing population in increasing economic security, he understandably felt a new confidence in his ability not merely to survive but to conquer any obstacles. Such assurance continued to blind him to the harm he was doing to the basis of his own existence. It has taken nearly two hundred years for this apparently unassailable confidence to begin to give way to the double realization that both man himself and the natural environment he has learned so effectively to exploit are terribly fragile. The reason of course is that they are truly interdependent. As man "conquers" or weakens the ecological context in which he exists, he weakens or destroys himself. In the longer run, man may prove to be more fragile or vulnerable than nature. It can exist without him; he cannot exist without it.

ECOLOGICAL BALANCES

Man has been hacking away at the forests with ever-increasing effectiveness since early Neolithic times, and in most areas has followed this procedure with agriculture. (Some of the dimensions and consequences of deforestation are examined in Chapter 9.) He has replaced one vegetational system, the product of many millennia of adjustment to climate, soils, water resources, fauna, and other elements acting in concert, with another which is a largely artificial construct. Most agricultural systems concentrate on the growing of plants originally derived from other and distant areas and refined by selective breeding over time so that they little resemble the original wild parent; most of them cannot maintain themselves in competition with wild species. Natural landscapes are occupied by a series of plant communities, a variety of species sharing a given environment in balance with one another and with the local fauna, all in adjustment to the other environmental characteristics of the area.

Agriculture removes all this and substitutes for it a single plant species, at least in any one field. Some farming systems may grow several crops (although rarely in the same field), but others concentrate heavily or exclusively on a single crop, such as cotton, rice, corn, or wheat. The greater the departure from the original varied complex of plant communities, the greater the risk that the ecological balance may be seriously disrupted. Each plant has a different set of requirements for soil, moisture, temperature, and sunshine. To cover an area with a single species instead of the mix which originally occupied it is to impose unnaturally heavy demands, which usually have to be met in part by artificial means: irrigation, fertilization, and the elimination of competition through weeding. In the longer run, however, irrigation and fertilization cannot sustain productivity or soil qualities indefinitely. In an agricultural system, the plants and often also their residues are removed from the field instead of being allowed to complete the natural cycle of growth and decay, returning the decomposed materials to the soil and preserving both its fertility and its structure. This cannot adequately be compensated for with the use of chemical fertilizers, especially when the soil is supporting only a single plant species. Over a period of time irrigation tends to concentrate in the soil an accumulation of minerals carried in solution or may also lead to waterlogging. In arid and semi-arid areas, in particular where irrigation is most needed to sustain agriculture (and where it is therefore especially unnatural), the accumulation of minerals or waterlogging may relatively quickly lead to declining yields and eventually may make the soil wholly unproductive, in the absence of adequate rainfall to flush it or of natural drainage to keep it free.

Concentration on a single plant (or animal) species also greatly increases vulnerability to disease. Natural plant and animal communities are provided in effect with a series of fire breaks; most diseases are specific to particular species and cannot spread through other species, but multiply very rapidly in an area covered by a single crop plant. The same is true of most insect pests. Poison sprays to attempt to control such ravages have already been shown to present risks to human life which may be worse than the problems they are designed to prevent. The use of chemical fertilizers to maintain yields also adversely affects the nutritional qualities of the crop, and some of them may have a toxic effect. This discussion should not lead to the hopeless conclusion that deforestation and agriculture were all a terrible mistake and that we should return to a hunting-and-gathering culture. But it does make clear that alteration of the environment can be a two-edged sword and that it must be done with care and with some knowledge of the possible consequences. With careful management and an awareness of the fragility of the ecological balance, the worst mistakes can be avoided and the damage minimized.

The most destructive, and hence most fragile, forms of agriculture are those which concentrate exclusively on a single crop, which is grown year after year. Cotton is a good (or a bad) example, especially in terms of the history of many cotton-growing areas in the southeastern United States, whose soils were both exhausted and eroded to the point where agriculture in any form was no longer possible in large parts of many formerly productive cotton regions. The area's demise was also hastened by the disastrous spread of cotton diseases and insect pests, including the boll weevil, which found this artificially uniform environment an ideal hothouse. (Fortunately, irrigated rice, globally the most important crop grown on a monoculture basis, is less susceptible to disease, and paddies are protected from erosion by diking, terracing, and their cover of irrigation water, which also brings in plant nutrients, so that with the addition of careful fertilization with organic wastes, yields can be maintained over very long periods.) A better and more stable ecological balance can usually be achieved through the use of crop rotation, by growing several crops (in different fields) simultaneously and by combining annual field crops with tree crops and with animal husbandry, thereby also providing a source of manure. However, such a solution reduces rather than eliminates the problems which any agricultural system creates by constructing what is necessarily an artificial environment and hence disrupting the ecological balance. Any alteration of the environment carries the risk of consequences which may be more detrimental than beneficial, in the short or long run.

Artificial introduction of animal species may also have unforeseen and harmful consequences. The introduction of the rabbit into Australia is probably the most dramatic case, since in the absence of natural enemies it multiplied out of hand and severely damaged the environment for most other species (including man). Involuntary introductions, such as the spread of the Japanese beetle into North America in the 1930's, have been only slightly less destructive; the starling, referred to in Chapter 4, has become a major pest. Both species found in effect an ecological niche in North America not occupied by other species, and became successful enough to upset the balance so that they made severe inroads on other plant and animal life, including for the most part species and plants which were considered far more valuable or desirable. The use of imported species to fight what are regarded as pests is always dangerous because the results are unpredictable and may backfire. Hawaiian sugarcane growers some years ago introduced the Indian mongoose hoping that it would eliminate or reduce the rat population, which had mushroomed in the cane fields and was doing great damage. The mongoose instead proved to coexist more or less peacefully with the rats and became an almost equally important pest to the cane growers.

SEWERS IN THE SKY AND THE SEA

Agriculture at least attempted some degree of cooperation with and awareness of environmental conditions. Man's other activities, especially during the past century or so, represent a far more drastic departure from the concept of ecological balance and have for the most part been undertaken with no regard at all for the possible consequences, or in complete ignorance of them. The pollution of the atmosphere and of water bodies has become a nightmare as the result of industrialization and the attendant explosive increase in the burning of fossil fuels, the massive use of chemicals in manufacturing, nuclear bombs, defoliants, and the mushrooming of the internal combustion engine. Most people have become aware of these problems, but the solutions require more drastic change in human life styles than most people seem ready to adopt. Although we still know distressingly little about the dimensions, trends, and consequences of atmospheric and water pollution, we know enough to justify profound alarm.

Systematic study of this problem has begun at the eleventh hour, when it may already be too late. Atmospheric pollution affects a huge and complex system of air circulation and regeneration. We have been pouring gases and particulates into the air at an accelerating rate, especially since about 1900, and can be certain only that the effects are bad enough to warrant severe control measures. We do not yet know in any detail how and to what extent atmospheric pollution may affect weather and climate on a macro scale, nor the full scale of its effects either on plants and animals or on man himself, except that whatever affects one part of the ecological system affects man, who is an interdependent part of it, whether or not it directly gives him silicosis, emphysema, or cancer. The oceans also constitute a vast system of circulation, and a series of ecological systems with which man is involved as a user of oceanic resources. It has been startling to discover in recent years the extent to which even the major ocean basins have been polluted and altered by human action, a combination of haphazard waste disposal and over-fishing which has already changed fundamentally the entire marine system and which may have consequences nearly as serious as those of atmospheric pollution. Although we still know relatively little, mainly because study has begun so late, it is clear enough that pollution control presents no insuperable technological problems. The means are straightforward and technically well within our power. The problem is to persuade people, corporations, and governments to apply these means and to accept the higher costs and the inconveniences which would probably result in most cases. Most people find it difficult to make short-term sacrifices in order to maximize their long-term welfare. But in ecological terms, the hour is late and drastic measures are called for.

The burden of man's impact on his natural environment has been greatly increased not only by industrialization and the revolution in power use but by the rise in human population which has resulted from the increase in productivity. This problem is discussed in greater detail in Chapter 13, but it needs to be mentioned here as perhaps the most basic aspect of the ecological revolution. Man has upset the ecological balance primarily by multiplying his own numbers and by spreading his occupance and use of the environment over almost the whole of the world, while at the same time increasing the variety of his demands and impact on the ecological system. Although increased human survival and greater numbers may be or may seem a strongly positive matter, it clearly has negative consequences also as it further strains or subverts a total system on which man is just as dependent as are its other components, especially as he continues to weaken those other components. Finally, one may raise the more philosophical question of whether man can be preserved at all unless he can learn not to destroy himself, indirectly by destroying his environment or directly through war and hate. Man, the supposed conqueror of nature, has much to learn. Perhaps it is the beginning of wisdom that he has at last come to recognize some of his own limitations.

CHANGE AND PROCESS

Fortunately, neither man nor his environment is static. The ecological system is best seen as *process* rather than as a fixed set. As part of this system, man changes too and will continue to change, hopefully in a more effective response to his ecological setting than heretofore. The environment as a whole, and each part of it, is continually changing and adapting in a mutually linked system. Diurnal, monthly, and seasonal alterations of this sort are obvious, but on a much larger time scale also the environment can be seen as dynamic. Soils and vegetational complexes, for example, alter slowly over time in a mutually responsive fashion, and also in response to changes in climate. It is possible to reconstruct these changes throughout most of the earth's history by the use of fossilized evidence of fauna, flora, and major geologic epochs and to measure their magnitude. Man appeared on the scene extremely recently on such a time scale and his long-run place in the ecological system is therefore still unclear; no doubt much will depend on how well he is able to adapt.

Developed particularly in the past thirty years, analysis of plant pollens preserved in the soil is a technique which makes it possible to pinpoint with considerable accuracy even short-term and minor changes in vegetation, and hence in climate during this brief period in which man has been present. Pollen grains are extremely well protected against decomposition and survive in the soil under most conditions for twenty or more millennia. This power of survival extends their evidence in time well into the period of the last glaciation. As pollen from plant species which tolerate cold, drought, dampness, or heat vary in their occurrence and mix in a given area over time (temporally sorted by their vertical stratification), one can see climatic changes mirrored over periods as short as a few centuries. During the evolution of the human species, man has had to adapt to drastic changes in climate and in the other elements of the environment. The emergence of homo sapiens, on the order of fifty thousand years ago, coincided with the last advance of the ice, and he began to evolve toward agriculture, domestication of animals, and ultimately civilization during the equally drastic changes accompanying the subsequent retreat of the ice sheets and the return of warmer climate. Man has so far thus proved himself eminently adjustable and has taken maximum advantage from both major and minor environmental fluctuations rather than being limited or eliminated by them, as has been the case with many other species. His greatest challenge, however, is of his own making: having reduced the one-way power of the environment over him, can he now continue to manipulate it without destroying its ability to sustain him? Can he, in other words, change now to adapt to the new problems which he has himself created by upsetting the ecological balance? Or will he be eliminated, as so many other species have been, by his inability to change, to reverse his present disastrous course?

Much of the answer will depend on how man perceives his environment: as obstacle, as enemy, as tool, or as the basis of his existence which requires nurturing care. Culture forms a lens through which people perceive reality. Since culture changes over time, and is different in different places at the same time, the distortion of this lens is not constant. It is perfectly possible, in other words, for man to alter his perception of the environment in the face of this new challenge so as to adapt his behavior accordingly. We may already be seeing the beginnings of such a change, as people in a variety of contemporary cultures become more sharply aware of what are in fact relatively long-standing problems such as atmospheric pollution, to appreciate their urgency, and to be ready to do something about them. General standards of what constitutes good or acceptable or clean air have changed radically and rapidly since 1900, since 1950, and since 1970 in this country as the atmosphere has been progressively degraded and the problem has consequently forced itself onto the attention of more and more people. The same phenomenon has been repeated in western Europe, the Soviet Union, and Japan as industrialization and the spread of the internal combustion engine have acquired similar scale. Tokyo, for example, with probably the most polluted air of any big city in the world, may also be the most pollution-conscious. In other words its citizens may perceive atmospheric pollution especially sharply. At major street junctions in the downtown area automatic pollution sensors have been mounted which flash continuous readings of the changing count on huge illuminated billboards.

PERCEPTION OF HAZARDS

Even in Japan, however, action has been slow. Tokyo continues to grow, and private car ownership and use continue to multiply. The situation is comparable to human behavior in response to other environmental hazards, even those which are clearly perceived. Although the environment as a whole changes relatively slowly over time, periodic extremes may occur every few years, such as drought, flood, earthquake, volcanic eruption, and (one must now add) concentrated as opposed to chronic smog. These and other such dramatic events are well recorded and reported. They can be charted over time, and hence to some extent predicted, at least in the sense that they can confidently be expected to recur within a certain period, and in certain well-defined areas. In general, people tend to disregard this by adjusting their behavior to the usual rather than to the unusual. Settlement has not by any means avoided the San Andreas fault area in California despite the disastrous earthquake of 1906 and the frequent reminders that another major earthquake will recur there. The slopes of Mt. Vesuvius are settled and farmed. About an eighth of the U.S. population lives in areas subject to periodic flooding; a similar proportion lives in areas subject to periods of major drought, although drought is more difficult to define and may not so immediately affect the welfare of non-farmers. At least one-third of the U. S. population lives in areas of chronic and periodically acute smog. Perception of environmental hazard is obscured to some degree, although as pointed out above it is changing. In general, people continue to pay more attention to what they perceive as the rewards of living in environmentally hazardous areas than they do to the risks, and they discount the unusual, suppressing or perceiving only vaguely the certainty that the unusual will happen again, and probably many times in their lifetime, since it is the nature of the environment to fluctuate.

SMOG

Acute smog builds up under certain atmospheric conditions of temperature, sunlight, and air movement (or its absence). Sunlight in particular acts on the gases emitted from automobile exhausts and certain other industrial emissions to produce ozone, a toxic gas. In the absence of air movement, or when what is called a temperature inversion exists (a layer of colder air trapped below a layer of warmer air so that the whole system is static instead of rising), toxic concentrations build up instead of being dissipated. This is the common pattern in the Los Angeles area, which lies in a coastal basin surrounded by mountains, and in other parts of the coast above a cold ocean current, and massive car and industrial emissions in relatively enclosed basins combine to produce both chronic and acute smog. Certain plant species, especially cultivated forms such as lettuce, may show damage first, but effects on the human population may be equally great, at least in the long run, and far more serious than the eye-watering and coughing which are short-run symptoms of irritation.

Conditions conducive to acute smog are fortunately not continuous; temperature inversions are more uncommon than common even in the Los Angeles basin, and there is usually some air movement. Again, people have for the most part perceived or adjusted to the usual rather than to the unusual, as they have done in other areas where acute smog is a recurrent and even lethal phenomenon, mainly in other enclosed or semi-enclosed basins where there is a low, dense cloud layer and a high concentration of industrial, household, and automobile emissions. The London basin is a famous example, but to a degree also an encouraging one in the sense that the smog hazard was perceived and counter measures taken to restrict emissions. Most of the London urban area was declared a "smoke-free" zone, and although the law was not applied literally, emissions were greatly reduced and acute smog became less frequent. Unfortunately it had to kill or cripple a great many people before the hazard was seen clearly and urgently enough to produce some action. Unfortunately also, the common response to dramatic.disaster is temporary· chronic problems are not so easily perceived nor is behavior so easily attuned to the risk they present.

CAN MAN SURVIVE?

Fortunately, perceptions and behavior change. In a world where there is beginning to be nearly global exchange of information and awareness of differences in experience and culture, each area may also be able to benefit from the perspective, insights, and mistakes of others. All mankind shares a planet whose environment operates as a single system. What man does to it in one area is likely to affect its nature elsewhere. Nuclear testing in the south Pacific or central Asia puts radioactive particles into the system of global atmospheric circulation and into the similar circulation system of the oceans so that all parts of the world are covered. Industrial pollution in any area also enters the global circulation system. The effects of biological and chemical warfare in Indochina cannot be confined to that unhappy area. We are all in the same boat, even if some of the most obvious leaks are only in certain parts of it. The more industrialized nations have necessarily the major responsibility, partly because they are the chief polluters and also the chief users of finite resources, partly because their degrading effects on the environment are inherited by everyone else, the poor two-thirds of mankind, whose economic margin is much smaller and who therefore can less well tolerate the consequences of a deteriorating environmental base. The United States alone, with less than 7 per cent of world population, accounts for over one-third of world consumption of fossil fuels and generates nearly half of the world mechanical energy, producing in the process nearly half of the world's atmospheric and water pollution. This is both irresponsible and short-sighted and will be increasingly resented. Change is urgently overdue. All mankind is interdependent, just as man in turn is interdependent with his environment.

Something may be learned from the experience of other cultures, if only the realization that perception of the environment and attitudes toward it have varied greatly from place to place and from period to period. Most of traditional Asia, for example, saw man as an integral part of a cosmos dominated by nature. Contentment as well as material success could come only through acceptance of the rightness of man's adjusting himself to the greater natural world of which he was a part. This attitude was to some degree common to many agrarian societies elsewhere, but in China especially it was enshrined as a central part of a philosophical and moral system. The bulk of imperial administration was devoted to the care of the land. The elite and the peasants alike evidenced a deep respect, almost a reverence, for a natural order conceived as grander than man and more to be admired. The Chinese certainly altered nature, through terracing, irrigation, deforestation, and intensive cultivation of an immense landscape. But such activities were not seen as pitting man against nature but as carrying out the role of a careful and respectful steward. Cooperation with nature was necessary; it could benefit man only if he accepted the limits it was seen to impose. With this kind of care, the Chinese environment became without question the most productive agricultural area in the world, and by far the most consistent, continuing to produce high yields for over two thousand years. Admittedly, large-scale industrialization was lacking and this was therefore largely an agriculturist's view.

The Western attitude toward nature, though more varied, has tended instead to see nature as an antagonist and man as its conqueror. Even long before the Industrial Revolution the West enthroned man as God's supreme handiwork, and tended to assume that he had in fact a duty to subdue nature. Only by doing so could he achieve the "progress" in which the West also believed. Where in the traditional Chinese view mining was devalued and even considered impiety because it robbed the earth instead of making it more productive, as agriculture was seen to do, Western thinking emphasized man's mission to wrest nature's secrets from the earth and to exploit it. As industrialization progressed, metals and minerals were seen, through the machines they made possible, as "freeing" man from his earlier dependence on nature and even as cancelling out the importance of the environment as a factor in man's affairs. With the advent of industrialization in Asia, and of an originally Western ideology (Marxism-Leninism) in China, many of these Western attitudes toward the environment have spread eastward (see the reading list for this chapter). To a great extent, of course, industrialization is not culture-bound and brings about a fundamental change in attitudes and behavior wherever it spreads and whatever its cultural origins or adopted context. But some of the traditional Asian attitudes remain, even in highly industrialized and urbanized Japan or in Communist China and developing India. Perhaps Asians will be able, with the help of their traditional values and perceptions, to make a less disastrous accommodation between industrialization and the natural environment than the West, with its different traditions and perceptions, has managed to do.

In Japan and China especially there are already some signs that this accommodation may be beginning to happen in the form of controls and in the form of regional planning for dispersal. Technologically also we still have much to learn about the consequences of and the means for controlling environmental degradation. New technology is continually being evolved; the directions it takes are determined in large part by felt needs: what a given society considers important or urgent, which is in turn dependent on that society's perceptions. There may well be a sharper perception of the environment, and of the urgent need to limit its degradation, in parts of Asia than elsewhere in the world. Having pioneered industrialization, the West may well now have to learn from the East how to make industrialization viable and livable. One of the salient characteristics of Asian society has alway been its emphasis on the general welfare as opposed to individual license. This emphasis has been most consistently and strongly manifest in the family system, but it has also shaped behavior and policy in the large. Contemporary China is a clear example, but the tradition of individual subservience to public good is in fact very old in China and in most other Asian cultures.

The West, especially the United States, has tended instead to emphasize individual freedom—"rugged individualism." In the frontier past, while space and economic opportunity were ample, such an attitude was understandable and viable. In a more crowded world where individual actions profoundly affect public welfare, they may be less so. In any case, environmental degradation can be checked only by concerted action, something for which there is old and varied precedent in many Asian societies. Perhaps they can lead the way or perhaps the West can alter its own perceptions in its own way and can begin before it is genuinely too late to reverse the disastrous course which industrialization has set. We live in a closed ecological system; it can nourish or extinguish us; the choice is ours.

Questions for Further Study and Discussion (Chapters 4 and 5)

1. Many cities on or near physical boundaries or transition zones are relatively small, and others far from such boundaries or zones are very large. How do you account for this? Find examples of both.

2. What examples can you find, in addition to those cited, of similar environments which are used differently? Of different environments which are used similarly? How can each case be accounted for?

3. What other examples can you find of the similar use of similar environments which help to support the idea that the physical environment does influence man in a more or less consistent association? To begin with, consult a series of atlas maps dealing with world agricultural and crop patterns, but do not limit your investigations to agriculture.

4. How in particular cases may environment influence culture? How may culture act upon environment?

5. How does the physical environment affect regional exchange or spatial interaction?

6. What environmental factors are relevant in the site of the city where you live? In its economic specializations? In its cultural characteristics? How have spatial relations been important in all of these matters?

7. Why is Trenton, at the falls of the Delaware, not the main break-in-bulk point, and why is it so much smaller than Philadelphia? What are Trenton's major functions, and why? (An atlas map or maps will be necessary for this and most of the following questions.)

8. Many ocean-going ships can reach Albany on the Hudson River. Why is New York the major break-in-bulk point and by far the bigger city?

9. Why is there no large city near the mouth of the Connecticut River? Why is there no large city near the mouth of the Columbia River?

10. Why are the largest cities of the fall line in eastern United States north of Richmond?

11. Why did Chicago become second in size only to New York among American cities? What various specific factors must be considered in explaining Chicago's growth?

12. During much of the colonial period, Boston was larger than New York. Why did New York overtake it, and why did Boston come to be relatively low on the list of major American cities in terms of size? In particular, what environmental and what spatial factors were important?

13. To what extent do the Swiss political frontiers coincide with the end of the mountains and the beginning of plains or broad valleys? With physical lines such as rivers or mountain crests? How do you account for the discrepancies?

14. Compare Switzerland and Bhutan. Their physical environments are similar, and the countries are about the same size. In what ways are they different, and why? In what ways are their cultures and economies similar and why?

15. How does Switzerland rank among the countries of the world in per capita income? What do you conclude from this about the effect of environment on the Swiss economy?

Selected Samples for Further Reading (Chapters 4 and 5)

Anderson, W., ed. *Politics and Environment: A Reader in Ecological Crisis.* Chicago, 1970. A collection of recent articles dealing with environmental pollution and some efforts—and non efforts—to combat it.

Bach, W. *Atmospheric Pollution.* New York, 1972. A brief clear survey of a complex and rapidly developing field.

Bates, M. *The Forest and the Sea: A Look at the Economy of Nature and the Ecology of Man.* New York, 1960. A beautifully written discussion in essay form of the interrelation between man and nature, in the context of the biotic community as a single whole.

Brown, L. R. *Seeds of Change: The Green Revolution and Development in the 1970's.* New York, 1970. A summary treatment of the development of high-yielding varieties of major cereals, and the promise they hold for food production, and the problems they pose for equity.

Brunhes, J. *Human Geography.* Abridged ed. Trans. E. F. Bow. London, 1952. A classic French view of geography largely as a study of man's adjustment to his physical environment, but from the possibilist point of view.

Burton, I., and Kates, R. W. "The Floodplain and the Seashore," *The Geographical Review*, LIV (1964), 366–85. Hazard perception and behavior by residents of areas exposed to flood risk.

Butzer, K. W. *Environment and Archeology.* Chicago, 1964. An authoritative geographical synthesis of the environmental conditions of the Pleistocene, and the uses of this material for the analysis of man-land interrelations in prehistory.

Chorley, R., ed. *Water, Earth, and Man.* London, 1969. A good recent general text in physical geography, by a variety of specialist authors, designed for the beginning student.

———, and Kennedy, B. A. *Physical Geography: A Systems Approach.* Englewood Cliffs, N. J., 1971. A somewhat technical and highly quantitative text, with extensive bibliography, which emphasizes systems analysis.

Chute, R. M. *Environmental Insight.* New York, 1971. A biologist's view of the scope and nature of environmental change, especially as affected by man.

Dansereau, P. *Biogeography: An Ecological Approach.* New York, 1957. A well-written survey by a botanist of the origins and spread of useful plants and animals, the influences of climate on biotic communities and their cycles, and man's impact on the natural landscape.

Detwyler, T., ed. *Man's Impact on Environment.* New York, 1971. A well selected compendium of articles by a variety of specialists, with integrating essays by Detwyler, on the wide range of destructive effects man has produced in his environment, and a balanced set of appraisals of the problem as a whole.

———, and Marcus, M. *Urbanization and Environment: The Physical Geography of the City*. Boston, 1972. The city as a special sort of eco-system, where man's impact on nature is concentrated most overwhelmingly; chapters by several specialists are included.

Ehrlich, P. R., and Ehrlich, A. H. *Population, Resources, Environment: Issues in Human Ecology*. New York, 1970. A useful general discussion.

Eyre, S. R., and Jones, G. R. J., eds. *Geography as Human Ecology*. New York, 1966. Eleven essays on an old and now newly popular theme.

Finch, V. C., et al. *Elements of Geography: Physical and Cultural*. 4th ed. New York, 1957. For many years a standard text in physical geography, including also a briefer treatment of some of the aspects of human settlement.

Firth, R. *Malay Fishermen: Their Peasant Economy*. London, 1946. An anthropologist's study which places a social group in its interrelated environmental and cultural context.

Fonaroff, L. S. "Malaria Geography," *The Professional Geographer*, XV (1963), 1–7. An example of the interplay between man and the biotic environment, with a great number of important and fascinating implications.

———. "Man and Malaria in Trinidad: Ecological Perspectives of a Changing Health Hazard," *Annals of the Association of American Geographers*, LVIII (1968), 526–56. A detailed sample study of the relationship between culture, environment, and disease.

Forde, C. D. *Habitat, Economy, and Society*. London, 1934. An investigation of the relation between physical environment and human culture, including several sample studies and an excellent general discussion.

Glacken, C. J. *Traces on the Rhodian Shore*. Berkeley, 1967. A masterful and fascinating survey of western philosophers' changing ideas about man's relation to his environment and his impact on it, from classical times to the eighteenth century, with an extensive bibliography.

Goldman, I. *The Cuebo Indians of the Northwest Amazon*. Urbana, Ill., 1963. A study of cultural change in an isolated area with a demanding physical environment, and of the interaction between culture and environment.

Greenwood, N. H. *Human Environments and Natural Systems*. Boston, 1972. An introductory text in the booming field of ecological studies, which emphasizes man's conflict with his environment.

Jones, E. "Cause and Effect in Human Geography," *Annals of the Association of American Geographers*, XLVI (1956), 369–77. A discussion of "laws" in social science, with the particular example of environmental causation.

Laporte, L. F., et al. *The Earth and Human Affairs*. San Francisco, 1972. A multi-author survey by the Committee on Geological Sciences of the National Academy of Sciences which gives a critical historical overview of man-environment interrelations and also an introduction to environmental processes.

Leeds, A., and Vayda, A. D., eds. *Man, Culture, and Animals in Human Ecological Adjustments.* Washington, 1965. Case studies of the relation between men, animals, and plants in a variety of cultural and environmental settings.

Lewallen, J. *Ecology of Devastation: Indochina.* Baltimore, 1971. How war has affected the Vietnamese environment: a deeply disturbing account.

Loewenthal, D. et al. *Environmental Perception and Behavior.* Chicago, 1967. How different cultures view their environments, and how these perceptions influence behavior.

Matley, I. "The Marxist Approach to the Geographical Environment," *Annals of the Association of American Geographers,* LVI (1966), 97–111. How a powerful ideology affects environmental perception.

Murphey, R. "City and Countryside as Ideological Issues: India and China," *Comparative Studies in Society and History,* XIV (1972), 250–67; and ibid., "Man and Nature in China," *Modern Asian Studies,* I (1967), 313–333. Two essays on the Asian perception of the natural environment and (in the Chinese case) the change in attitudes as a result of revolution.

Saarinen, T. F. *Perception of Environment.* Washington, 1969. A brief, clear survey of a booming field, published by the Association of American Geographers.

Semple, E. C. *Influences of Geographical Environment.* New York, 1911. A persuasive and vigorous statement of environmental determinism, based on the earlier German school, now discredited and rejected but still well worth reading with a critical mind.

Singer, S. F., ed., *Global Effects of Environmental Pollution.* New York, 1970. A collection of studies by specialists.

Spate, O. H. K. "The End of an Old Song?" The Determinism-Possibilism Problem," *Geographical Review,* XLVIII (1958), 280–82. A critical review of recent literature on environmental causation and some suggested conclusions.

———. "Toynbee and Huntington—A Study in Determinism," *Geographical Journal,* CXVIII (1952), 406–28. Arnold Toynbee's provocative many-volume work, *A Study of History,* and Ellsworth Huntington's many publications on the climatic influence are subjected to a devastating criticism.

Sprout, H. and M. *Man-Milieu Relationship Hypotheses in the Context of International Politics.* Princeton, 1956. A perceptive analysis of the interaction between human society and the natural environment (milieu) as applied to political phenomena.

———. *The Ecological Perspective on Human Affairs.* Princeton, 1965. An imaginative analysis of the interaction between human society and environment, as applied to political affairs.

Strahler, A. N. *Physical Geography.* 3rd ed. New York, 1969. A recently revised standard text, well balanced in its treatment of the physical environment.

———. *Introduction to Physical Geography.* Rev. ed. New York, 1965. A shorter version of Strahler's earlier standard text, emphasizing the aspects of the physical environment most relevant to human society.

Thomas, W. L., ed. *Man's Role in Changing the Face of the Earth.* 1956. A large volume representing the collected papers and discussion presented at a symposium whose broad purpose is stated in the title and which includes a number of important and thoughtful articles.

Tuan, Y. F. "Discrepancies Between Environmental Attitudes and Behavior: Examples from Europe and China," *The Canadian Geographer*, XII (1968), 176–91.

———. *Man and Nature*, Washington, 1971. A brief speculative overview of changing human attitudes and behavior toward the natural world, published by the A. A. G.

Van Dyne, G. M., ed. *The Ecosystem Concept in Natural Resource Management.* New York, 1969. A good collection of essays.

Van Riper, J. E. *Man's Physical World.* 2nd ed. New York, 1971. A good standard text, comprehensive and up to date.

Wagner, R. H. *Environment and Man.* New York, 1971. A useful textbook which is also an impassioned tract for better management of the environment.

Webb, W. P. "Geographical-Historical Concepts in American History," *Annals of the Association of American Geographers*, L (1960), 85–97. A stimulating personal account of how the author, a historian, became interested in geography and especially in the environmental influence, and of the importance he attached to this approach in his classic study of the North American westward movement, *The Great Plains.*

Whittaker, R. A. *Communities and Ecosystems.* New York, 1970. A somewhat trendy but generally sound survey of the new field known as "ecology."

Wilson, C. M., and Matthews, W. H. *Man's Impact on the Global Environment.* Cambridge, Mass., 1970. A stock-taking account by earth scientists working with the Study of Critical Environmental Problems (SCEP).

6
The Region

The third function of geography has been listed as a study of "the regional framework and the analysis of specific regions." Examination of spatial form or of the spatial dimensions of human society must take account of distinct areal units and is therefore concerned with the concept of the region. The world may be divided into continents or countries or counties according to the existing more or less arbitrary lines. But there is a more important and more complex set of areal divisions to be made on the basis of units which contain distinct and internally consistent patterns of physical features or of human development.

ASPECTS OF AREAL DIFFERENCE

Such division is often difficult. There are almost endless criteria on which to base the definition or delimitation of a given region. An area may be distinctive and internally consistent because it coincides with the distribution of a particular type of climate, landforms, or soil, or because it is characterized by a particular type of economic activity, occupational structure, land use, or language. Areal distinctions may also be drawn on the basis of administrative or commercial services or trade relations which cover a particular territory. One may describe and mark off on the map manufacturing regions, agricultural regions, climatic regions, landform regions, religious regions, or political regions, to mention only a few of the more obvious ones. Few of these will often coincide exactly with each other, although in many cases the various factors or characteristics which make an area distinct may co-vary in their spatial distribution. The city as it is discussed in Chapter 2 is a good illustration of this problem. A city's newspapers will cover one area, its trade another (or a variety of others), the plain or river valley in which the city lies another, the city's specialized services, such as bus routes or library circulation, another, and its political control still another, the last one seldom coinciding with the compact urban area of the city.

For a variety of reasons it is essential to pick out some degree of regional distinction and regional coherence from the jumbled mass of distribution and areal differences. This is necessary in order to understand any one area adequately, to distinguish it from other areas, and to see the basic spatial framework within which man lives and which he has helped to create. Planning for the management and development of any area requires an understanding of the limits of the regional unit involved and of the region's distinctive components. Regional planning, in which development is planned for an area as a whole in all of its interrelated aspects, has come to be accepted as an essential method for promoting economic growth or areal development. It must rest on a knowledge not only of regional differences and their delimitation but of the multiplicity of interrelated factors, from climate to transport patterns to historical trends, which compose any regional unit. The businessman and the political administrator also need to delimit areas and to understand them for the efficient operation of their affairs and the subdivision or centralization of their activities. Trade areas and political boundaries are often matters of dispute, reflecting the complex problems involved, and the geographer attempts to provide some guidance. The study of political boundaries, what makes a good one or a bad one, where in particular cases the boundary should best be put, and what impact a political boundary may have on the area through which it runs, is a whole field in itself, within political geography, but it is also part of the geographer's more general regional function.

LINES, BARRIERS, AND TRANSITION ZONES

To make regional divisions intelligently, it is necessary first to understand as well as possible what gives the region its coherence and distinction, what its essential unifying or common qualities are, and secondly to determine the points or zones where these qualities significantly diminish and are replaced by another set of qualities belonging presumably to another region. Usually there is a wide zone of transition, and the line is difficult to draw. Prominent physical features such as mountain ranges often coincide areally with the zone of transition from one set of regional characteristics to another. Mountains may act as barriers to economic and cultural movement or interchange, and they may also be responsible for a transition in the physical aspects of the landscape. But the effect on mankind of even pronounced features of the environment is seldom constant. If such features are used as regional lines, the line may become obsolete or awkward as human occupation changes on both sides of such a boundary or meets across it. Even at a given moment in time sharp cultural gradients on the map are rare, and even in physical terms it is seldom possible to draw a neat line between regions. Regional characteristics change more or less gradually over space, and within the transition area between one region and another there may be a wide zone of ambivalence. This may be true even where a major mountain range is used as a regional line. The sea makes a clearer delimitation, but its barrier character has been tremendously lessened by the modern revolution in communications and transport. The North Atlantic Community, for example, now expresses in many respects more genuine unity among its members than exists among the countries of Europe as a whole, although the Atlantic is as wide and deep as ever. Rivers may be physical barriers to an extent, but where they are navigable they are usually unifiers in the sense that settlements on both banks are likely to be dependent on river-borne trade and hence will be part of a single economic and cultural community. Unnavigable rivers may support in their valleys a more or less uniform pattern of agricultural land use, which may be dependent on river water for irrigation. River basins are usually physical and human regional units. Attempts to divide them or to use rivers as political boundaries, as is commonly done, are unlikely to be satisfactory.

DISTINCTION, IDENTITY, AND REGIONALISM

Similar problems are involved in the drawing of almost any regional line, since regions tend to merge with each other over a wide zone of transition despite physical barriers. Any single line is thus not only arbitrary but likely to be seriously misleading as well as awkward, even though for most purposes and especially for political boundaries a single line must be used. Physical features, especially rivers, provide convenient lines and are therefore used despite the disadvantages. But for regional analysis, the basic problem is the determination of what gives any region its essential distinctive quality, and it is only on that basis that satisfactory regional lines can be drawn. Here we must be concerned with the physical framework, with economic, political, and social characteristics and functions, and also with the regional images and loyalties of the inhabitants. We are dealing in many ways with what psychologists call a *Gestalt,* a pattern of factors which tends to operate as a unit and which is more than the simple sum of its parts, as water is more than the separate totals of its hydrogen and oxygen components. Each aspect of an area exists in context, is interrelated with other aspects, and cannot be understood completely unless it is seen in its total context, nor can the area as a whole be understood unless all of its interrelated aspects are seen. The consciousness of this greater whole, of this areal pattern of characteristics, in the minds of the inhabitants is called "regionalism," or in the case of a national state, "nationalism." Separatism, or the desire for political autonomy, is an aspect of pronounced regionalism which is akin to nationalism but which may not insist on complete independence. Outside threats, pressures, or competition, real or imaginary, commonly strengthen regional consciousness and cohesion, but regionalism in basic terms results simply from the awareness of areal differences. People identify themselves with their own area and with its characteristics, including diets, dress, house styles, art forms, political attitudes, landscapes, seasons, and a host of other attributes which differ from region to region. Parents inculcate regional cultural traits in their children, take pride in their regional identity, and feel themselves to be different from and usually better than the people of other regions.

Such attitudes provide one of the clearest clues to the existence of regional entities and are particularly helpful when it comes to drawing regional lines. This may be done in some situations merely by asking the inhabitants to which cultural region they feel they belong. Regionalism is often especially pronounced in isolated or mountainous areas, where the physical distinction from surrounding areas is strong and where the culture and economy are likely to be distinctive also. Switzerland is a good example, and there are many others. But regionalism is not the particular property of the mountains, and every area which can be distinguished from neighboring areas feels it, in differing degrees. In the United States this feeling was strong enough in the southern states to support a movement for independence as a separate sovereignty, and it remains strong, partly because of sentiment and loyalties, partly because of the geographical differences which fostered the feeling of distinctiveness in the first place.

Language is often used as a measure of regionalism and as a basis for delimiting regions. Of all aspects of culture, language may be the most important mark of identity as well as one of the clearest, and it often does tend to coincide areally with most other aspects of a regional *Gestalt*. But although it may in many cases be a badge of regional identity whose importance is emotional as well as practical, language is not a necessary or sufficient cause for regionalism or an adequate measure of a regional entity. The example of the American Confederacy suggests that regionalism may be very strong and that a true region may exist without a regional language. Such examples could easily be multiplied. Switzerland, on the other hand, suggests that regionalism on other grounds than language may be strong enough to thrive despite language differences within the region. Until about the eighteenth century, language differences were seldom an issue or a cause for separatism, and multilingual states were the rule. People were not closely involved with one another over large distances, so that language differences were less important, and the literacy level was relatively low. In general, it was only with the rise of modern nationalism and the national state on an enlarged areal base that language became an emotional part of regional identity. In some cases, regionalist or separatist or nationalist movements have actually created or revived a regional language in order to emphasize identity; this has been especially true where the regional language was primarily a spoken one and had little or no literature—a literature would be manufactured. This certainly suggests that factors other than language are important in promoting regionalism and in creating a regional entity. Language has become a more important part of regional identity than it used to be, but it has not been primarily responsible for creating or establishing that identity. Improvements in communication and the increase of interaction in the last several centuries have also been accompanied by an increase rather than a decrease in regional consciousness and rivalry. As regions are involved more closely with each other, differences are perhaps more apparent.

It should be made clear, however, that the concept of the region is primarily an intellectual device, useful as a tool for the better understanding and analysis of human society. The regional concept picks out certain spatial patterns, but even political boundaries or other regional distinctions we may draw are products of our own minds rather than complete descriptions of reality, or of any "natural" order. In this sense, there are no "natural" regions, but only man-made or man-perceived classifications of the welter of areal differences.

THE REGIONAL HIERARCHY

The differentiation of regions may begin very low on the size scale, and it is important to establish the order or magnitude in which one is dealing in a particular case. Differences which one may call regional exist, for instance, between different parts of a city, and at the other end of the scale they are apparent between continents and hemispheres. Two neighborhoods or two cities will, however, have much in common as part of a larger urban, provincial, national, or even continental region. There is thus an overlapping hierarchy of regions, depending on the criteria used. One may look at the world or at any part of it through the big end of the telescope to see gross general spatial patterns, or through the small end to see more detail and more minute subdivisions. Both devices are useful. But the fact of areal difference is universal, and is expressed in ramified details of culture, from diet to dress, house types, language, prejudices, tastes, occupations, skills, and social and political systems. The United States is composed not only of a hierarchy of political subregions in the form of wards, cities, counties, and states, but of a great variety of more broadly based subregions involving areal differences in land use, employment, investment, and production, voting habits, income, marketing patterns, cultural attitudes, speech, and a host of other characteristics which have spatial form and are discernible as spatial patterns. Some of these broader spatial patterns are commonly recognized in geographical expressions as "the South" or "New England," but each of these regions may be further subdivided not simply on the basis of state lines but in recognition of more complex variations from one part of the South or New England to another. At the other end of the size scale, it is clear that both of these American regions as a whole belong within the larger region of the United States and in turn of North America; at that level, or on a world scale, their similarities become more important than their differences.

BASES OF DISTINCTIVENESS

Regional differences may in part be related to differences in the physical environment, but only in part. Human society is a complex affair in which a great variety of factors operate. The regional analyst must attempt to consider all of these factors. An area covered by a given climate or soil may support a more or less consistent type of land use or culture, related in part to climate and soil, but related also to non-environmental factors and indeed often occurring in physically different areas. A region may derive much of its distinctiveness from its interconnections with other areas. The Netherlands, for example, is a distinctive region in part because of its widespread trade relations with the rest of the world. The Rhine River, which is partly responsible for the commercial prominence of the Netherlands as a break-in-bulk point at the river's mouth, is also responsible for considerable economic unity in the Rhine Valley. Although political boundaries cross the Rhine and in some stretches use it as a political line, a heavy flow of trade follows the river. It is possible to argue that the "Dutch" region or the "Rhine" region actually extends beyond the limits of the conventional political or physical lines and includes aspects of the area with which the region trades. This kind of analysis, however, brings us back to the essential problem that few aspects of a region are likely to co-vary exactly, or to cover the same area, and that the application of a regional label to any area is ambiguous until the criteria used are first made clear.

The Example of China

Take, for example, the case of China. China as a political state covers a vast territory, containing within it a great variety of physical and human differences. About half of this area, in terms of square miles, is occupied predominantly by people who are ethnically not Chinese, speak a non-Chinese language, and have a non-Chinese culture and economy. Over half of it is so dry, so cold, or so mountainous that it contrasts sharply with the relatively well-watered, warm plains where most of the Chinese people live. On the other hand, prominent aspects of Chinese culture and large blocks of Chinese people are found in several places in east Asia beyond China's political frontiers: Chinese written language and culture forms in Japan, Indo-China, and Korea, heavy Chinese settlement in Thailand, Malaya, and Java. Where is the true China region? The answer depends on the criteria chosen, or on those to which the greatest weight is given. Generalized lines may be drawn on the basis of single factors—around the area where Chinese-speaking people are in the majority, or where agriculture is the predominant occupation, or where the rainfall exceeds twenty inches a year. This may be a useful service, but it begs the major question. What makes China China, and where does it cease to be China?

One convenient solution is to plot on a map the areal distribution of all factors which could be considered relevant as measures of regional identity: consistent language, physical type, economy, climate, landforms, social and political institutions, political control, cultural forms, historical alignments and changes, and so on down the list of any region's characteristics, although these will be of different relative importance for different regions. The basic minimum area to be included as indisputably China would be the area where all of the factors designated as consistent Chinese characteristics were found. But none of these characteristics covers exactly the same total area, and somehow the margins must be disposed of. A decision must be made about which of the characteristics are most important. Areas where they are all absent can be eliminated; areas where only some of them are absent will have to be treated on the basis of the priority given to each characteristic as a measure of Chineseness. By such means one can ultimately arrive at a reasonable regional differentiation, but only after it has first been decided what makes the region distinct. The plotting of areas on a map as suggested above will usually help to suggest which are the critical factors, but it will also be necessary to make a thorough analysis of the regional *Gestalt*, in order to avoid an arbitrary decision and to understand adequately the proper relative importance of the factors which must be evaluated. This is obviously not a simple problem, for all the ease with which lines are often drawn or the glibness with which people talk about "natural" regions. In the case of China there is a strong lead because Chinese civilization has traditionally rested on agriculture. Where agriculture and the farmer stop on the map, there "China" has tended to weaken or disappear. In this case most of the essential regional characteristics co-vary with agriculture in their areal distribution. In the excessively dry, cold, or mountainous parts of political China, agriculture is greatly reduced or eliminated, and many of the other "China" characteristics are replaced by a clearly non-Chinese set of characteristics, although since 1950 the Communist government has fostered much cultural uniformity.

It is then logical to ask why Chinese political control extended so far beyond the limits reached by the other factors, and why political control was not extended over the areas in other parts of east Asia where Chinese people or Chinese culture did spread. This question is raised not in order to answer it, for that would be beyond the scope of this brief example, but to emphasize the implications of the regional concept and the insight which it may stimulate.

REGIONAL CORES

Associated with the fact that in most regions the essential characteristics gradually shade off as the regional margins are approached is the further common feature of regional cores or nuclei. Usually spatial patterns of human development tend to focus or to coalesce on a hub or axis. This may be a physical phenomenon such as a fertile plain, a river valley, a seacoast, or a crossroads of easy routes. Or it may be a cultural development such as a city, the center of a particular type of regionally dominant economy, or a center of political or military control. Most clearly marked regional cores have all of these characteristics. They coincide with the largest area of dense population, include the largest share of the region's economic activity and the political capital of the region, and tend to set the region's cultural pattern. Usually this core area is the first in point of time to develop the cultural characteristics of the region as a whole and it spreads its dominant influence over the surrounding area as far as the original physical and cultural conditions allow. At some point these conditions become different enough or far enough away so that they cannot be adapted by influences exerted from the core. This distant area will lie in a zone of transition or will be attracted and influenced more powerfully by another less distant and more congenial center and will belong to that center's region. Most regions seem to be given much of their coherence by comparatively well-defined cores around which their areal traits tend to cluster.

Prominent examples of regional cores include the London Basin and the Paris Basin. Both are the chief centers of population in their respective national regions. Both are the centers of political control and the major market centers. Both acted historically as nuclei for the development and spread of their respective regional cultures, and both are still the dominant nuclei in this respect. Transport lines and circulation patterns focus on both to a high degree. The pattern of circulation and the density and degree of interaction within the region may often be a leading part of or clue to regional identity and distinction. Regional lines may be drawn on the basis of orientation in circulation patterns to a particular regional core, or on the basis of common interaction. Beyond such a line, orientation may be predominantly to the core of another region, even though it may be distant, and to a different set of interaction partners, although there may be a zone of ambivalence or indifference lying between competing cores. London and Paris are both strong cores whose dominance extends throughout and to some extent beyond their respective national areas. But in both countries there are also subsidiary cores, most importantly the centers of heavy manufacturing which developed much later than London and Paris and which are areally separate, associated with mineral resources largely lacking in the London and Paris basins. These lesser centers are subsidiary or tributary to London or Paris in part because they are mainly single-factor cores rather than multiple or generic in their nature. In other regions or countries more evenly balanced competing cores may develop as, for example, in Germany (Berlin and the Ruhr), and to some degree in the United States (the middle Atlantic coast and the Great Lakes–Chicago district). As with the hierarchy of regions, however, there is a hierarchy of cores which may be distinguished at different orders of magnitude. Strictly speaking, a core performs some service or exerts some influence over the whole of the larger region to which it belongs. Manchester and the English Midlands are subsidiary to London, and Chicago and the Great Lakes district to New York, but these subsidiary cores do play a prominent national role in the provision of certain major goods and services. St. Louis, however, or Atlanta, although they are centers of settlement cores within small subregions of the United States, do not play a marked national role and hence are not cores in the larger regional sense; they are merely centers of population clusters and of certain local functions or local regional characteristics.

SPACE, TIME, AND CHANGE

The delineation and analysis of regions is undertaken by the geographer because it involves the distribution and pattern of things in earth space. He is concerned with matters whose occurrence can be mapped. But he is also concerned with the relationships between regions, with the association of areas, and with spatial interaction. The patterns which he maps are not static, nor are they self-sufficient. They are part of an integrated whole which it is just as important to fit together as it is to take apart. Before one can examine relationships within the whole, it is necessary to distinguish the parts. This involves not only separate forces and factors but the concerted interplay of the whole complex of factors which makes up each region and the areal associations which result.

Here again is the parallel between history and geography, for in marking off areas and considering their association with other areas, the geographer is classifying space as the historian classifies time into periods. Each period is associated with other periods and is part of a continuous time sequence, but each has a body of characteristics which makes it distinct. Although each historical period and each geographical area is to some degree unique, there are both broad and specific similarities and interrelations between periods and between areas which the historian and the geographer must consider in attempting to understand, relate, analyze, and explain any matter which they investigate. In another sense, historical time and geographical space are each unitary and indivisible. To mark off segments of an interrelated and unitary whole is to some extent unreal and arbitrary, however carefully it is done. A distinguished British historian has said, "... such is the unity of all history that anyone who endeavours to tell a piece of it must feel that his first sentence tears a seamless web."[1] The same could be said of regional analysis. But distinctions are necessary and useful in time as well as in space, and beyond the obvious category of centuries, which may be compared areally with political states or with continents, the historian has the same problem in drawing lines between periods as the geographer in drawing lines between regions. It is not enough simply to divide time into centuries or space into countries. Each distinguishable historical period, and each aspect of each period, may cover a different block of time, just as each region and each regional aspect may cover a different block of space. But the varied group of essential traits which make a period distinct must be agreed upon, however arbitrarily, and a line or zone drawn which reflects this larger distinctiveness. Historical periods merge with each other over a time of transition. Lines are difficult to draw or may be misleading; as with regions, some aspects of one period may extend far into another period. And as with the geographer, the historian's job does not end with delimiting periods, but includes the analysis of specific periods and their relationships with each other as part of a temporal whole.

The period of scientific inquiry in Europe in the eighteenth century, for example, can be fully understood only in terms of its connections with the "Age of Reason" in the seventeenth century, with a variety of other periods and other places, and in the knowledge of what was to follow in the nineteenth and twentieth centuries. The manufac-

---

[1]F. W. Maitland, *The History of English Law before the Time of Edward I* (Cambridge, Eng., 1895).

turing region of the Ruhr and the Rhine valley in Germany can be fully understood only in terms of its connections with markets elsewhere in Europe, iron ore sources in Sweden, technological methods originally developed in England, and political factors in nineteenth-century Germany. The period of scientific inquiry overlaps the eighteenth century, as the industrial belt of north Europe overlaps Germany. One could equally well construct different regions and different periods out of these same temporal and spatial materials, depending on what one considered important or was specifically concerned with: cultural traits, rise of representative political institutions, degree of urbanization, nature of agricultural land use, or changes in social structure. Despite such complexity, necessarily involved in analyzing the infinite complexity of reality, the fundamental problem is simple. History and geography between them supply the basic dimensions of time and space which are essential in every investigation of human society. Both disciplines deal with classification, but as a basis for analysis.

In drawing regional boundaries, analyzing regional units, and considering the association of areas, the time dimension is of critical importance. Regional boundaries are seldom constant, since they reflect a particular stage of development. As the aspects of a society which give it its distinctiveness change, the degree and nature of its spatial expression change also. This is the result not only of technological change, such as improvements in communication which can extend unity, but of other temporal changes in the basic nature of the community. Wholly new factors or new emphases may arise in the economic or political system, for example, which produce a different core, the exclusion of former areas, or the inclusion of new ones. Invasions, conquests, or migrations may distort or basically alter regions. Cultures change, are powerfully influenced by other areas, or may be imported. New forms of production may arise to dominate the economy or to revolutionize its nature. There tends, however, to be a certain generic harmony of the components of any region which may reassert itself, although perhaps in a somewhat different form, even as the components change. Only over relatively long periods or under special circumstances do regions vanish or wholly new ones arise. The physical environment is relatively constant, and even though its influences change as man changes, it helps to maintain a certain amount of regional continuity over time. Spatial relations are also manipulated and altered as man changes, but physical distances and absolute location remain constant. While the "China" region has remained relatively constant areally during the past two thousand years (despite recent expansion into Manchuria and Inner Mongolia), tremendous changes or outside influences in Europe and North America have destroyed most of the regions of two thousand years ago and have produced new regions in their places. The same temporal processes are at work, slowly or rapidly, in any current situation, continually changing the essential quality of every region and thus affecting the region's spatial limits.

Any fact of human geography, or any region, can be examined and explained adequately only by considering its historical evolution. The location, size, and nature of a settlement, for example, may be explained statically, in terms of particular present advantages of local site or relations with other places. But it must also be explained dynamically, in terms of historical growth and of the changing factors which have affected that growth—cultural and technical changes, varying political influences or decisions, the rise and fall of new states, or changes in spatial interaction; any or all of these may be essential for the understanding of the present status of the settlement in question. As the separate regions change, so the associations between regions change, and indeed alterations may come about within a given region wholly or largely as a result of changes in its associations with other regions. Contemporary China is a good example; many of China's characteristics are being rapidly altered as a result of new and closer relations with the West since the eighteenth century, and more recently with the Soviet Union. Mexico and Canada are profoundly affected in their regional development by the juxtaposition of the United States; changes within them reflect changes in their great neighbor, the more because the friction of distance is at a minimum. But Mexico and Canada are also associated with other more distant areas, with Spain and Great Britain most obviously, and in differing degrees with the rest of the world. The nature and force of these associations is intimately bound up with the human development of the two countries. The association of areas in its broadest sense is relevant at every stage of regional differentiation. A single region, whatever its size, and whether it is based on a single functional criterion or on a more general homogeneity of characteristics, is simply an association of smaller areas which have enough in common to warrant their being distinguished from other associations of areas. This region is then involved in a broader pattern of association with other similarly constituted regions.

SINGLE- AND MULTIPLE-FACTOR REGIONS

Although we may speak in general terms of China or Mexico or Canada as regions, such a label must of course grossly oversimplify the complex of culture and environment involved in such large areas. We can more accurately speak of *single-factor regions,* such as political, linguistic, or climatic areal units or we can outline *functional regions,* such as trade or administrative areal units. The United States is divided, for example, by the Federal Reserve Bank into twelve districts, each intended to include a recognizable and viable commercial-financial region grouped around one dominant urban financial center (see Figure 10 on page 104). The Sears Roebuck Company divides the country into five major and eleven minor merchandising districts, each with its own urban center for warehousing and service for the district as a whole; each of these districts offers a slightly different set of goods, reflecting differences in regional needs or tastes and consumer preferences. The division of the United States and Canada into states and provinces is an obvious example of political administrative regions. Broad patterns of climatic differences are recognized

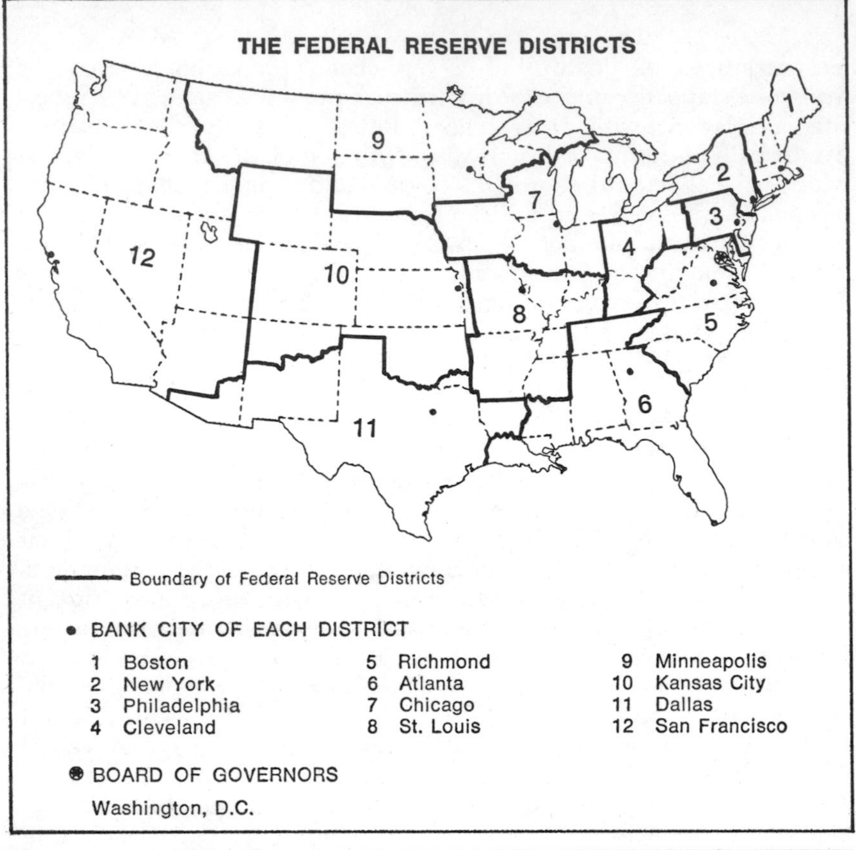

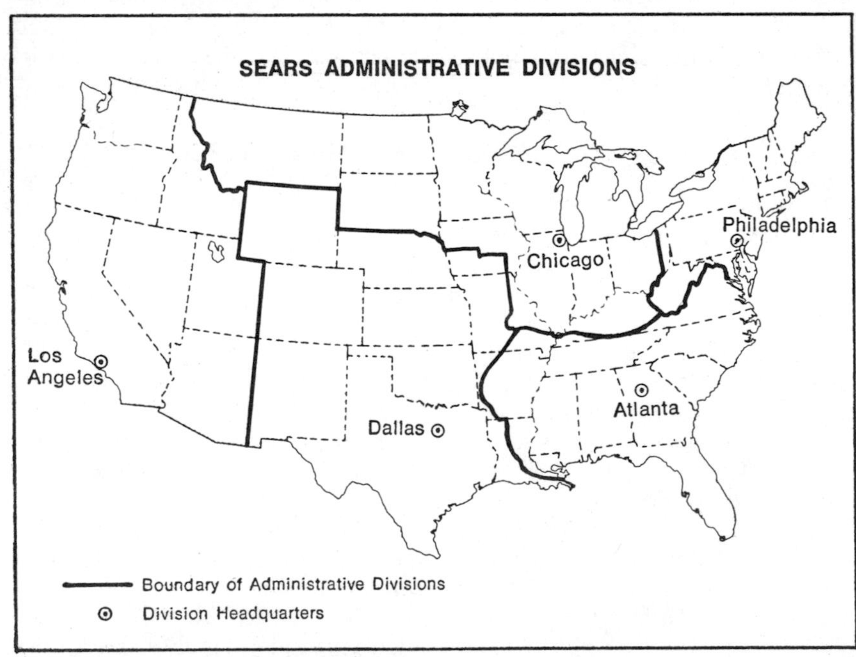

Figure 10. Sample Regionalizations of the United States.

in climatic regions. Single-factor regions such as these can of course be sharply delimited with a single line. A political boundary or the boundaries of a Federal Reserve District or of a precisely classified type of climate can easily be defined and will outline simply and clearly *one* consistent spatial pattern, extracted from the complex matrix of reality. But there is an undoubted unity implied and apparent in broader geographic expressions like "the South," "New England," "the Corn Belt," Mexico, China, or Canada, applied to very much more varied constructs including a great number of different characteristics. It may be difficult to define or delimit this unity, aside from the obvious political boundaries. Each of the single factors involved will almost certainly cover a different area, overlapping each other near the center but leaving a good deal of the "region" ambiguous. The American South, the Corn Belt, or New England, for example, are all multiple-factor regions whose identity depends on a variety of individual characteristics more strongly and unanimously present in some parts (Alabama, Iowa, Vermont) than in others (Kentucky, Ohio, Connecticut) which may lie spatially close to the edge of the region or may be influenced by adjacent regions. But the idea of the region assumes the existence of some spatial unit which exhibits a set of common traits. This broader concept of the region might be stated as a *definable area possessing a wide range of more or less homogeneously distributed and interrelated characteristics.* It is seen as a complex of homogeneous factors which are interdependent and which operate together as part of a larger whole, or *Gestalt.* The idea of the region has often been criticized because it is ambiguous or oversimplified. The distinction between single-factor or functional regions on the one hand, and broader, multifactor generic regions on the other hand, helps to avoid much of this criticism. The general concept of distinct areal homogeneity is extremely useful, provided it is based on an analysis of the many factors which make an area distinct.

STUDY OF REGIONAL WHOLES

Regional information and understanding are objectives of the geographer as well as the classification of regions. A large part of geographic work consists of the accumulation of regional material. Here also geography undertakes a task which is generally not done holistically, as an integrated whole, by other disciplines. The economist is interested in production, finance, or the market system in any region, the sociologist in kinship groups, social structures or institutions, and so on. The geographer attempts to examine and to understand the region as a whole, in all of its physical and human aspects, to fit these aspects together, and to relate the region to other regions. This is a demanding goal and may never be completely met, but someone must make the attempt to fit together the regional mosaic and to see regional wholes. So much has to be examined and understood in order to do this adequately that the geographer must depend on the specialized studies of others as well as on his own studies, and he frequently functions to some extent as an editor, or an integrator. He must also compile and analyze a great bulk of regional information. This collecting and editing is an essential but elementary first step, a means to an end. Other students of society need the information which geographers can provide. Regional planners, the general public, and those who make policy decisions of many kinds need regional information too. But the information alone is not fully useful until it can be fitted into the regional whole. It can also be dull to deal only with factual detail, to see only the individual trees and miss the pattern of the forest. Sometimes the "pack-rat mind," which collects endless bits of information for their own sake, is apparent in the craze for facts and in the confusion of information with knowledge which handicaps some regional study. Information can seldom be gathered intelligently unless an over-all pattern is seen. Buildings are constructed of bricks, but the bricklayer cannot, or should not, choose and lay his bricks without a blueprint or an image of the building. Knowing the number of miles of railroads in Chile, for example, is less important than knowing the patterns of distribution and interrelation which the railroads reflect and stimulate in Chile, the role of railroads in the regional economy, and the part which these patterns or factors may play within the whole of the Chilean regional unit. But the simple information is necessary first, and the need for that particular information is fully apparent only when the larger regional context is understood.

Holistic regional analysis attempts to delineate the particular use which man has made of the earth in a given region and the pattern and effect of spatial relations within it and between it and the rest of the world. Such analysis usually centers on population and its distribution, density, and growth, the nature of the settlements, the agricultural system, the use of other resources, the pattern and nature of manufacturing, trade, and transportation or circulation, and the nature and distribution of the regional culture as a whole. Consideration must also be given to historical factors and changes and to everything of significance about the region which has spatial form or spatial distribution. The regional analyst must in many cases depend on the work of particular specialists. Climate will be important to him, for example, but he will probably turn to climatological work for the detailed facts and processes, as he will in other cases turn to the work of soils scientists, agricultural economists, anthropologists, or historians. He must himself fit these varied data together within a regional framework which it is his business to provide. The regional whole must be understood as the context in which any of its parts are to be examined, since all of its parts are interrelated. A specialist's approach to agriculture, for example, might make much sense of the Russian agricultural system which a more general regional approach might miss; but it might also make much nonsense because the specialist did not adequately evaluate the influence of politics, tradition, or social organization on agriculture in this particular regional situation. Each region must be treated as a single areal manifestation of an integrated body of culture.

This kind of regional understanding is the basis for regional comparisons, which may throw light on whatever is being examined, whether it is a single aspect of the region or the whole complex. Comparison is obviously necessary in making regional distinctions. Historical comparison is necessary, too, for any region is a dynamic entity whose nature is continually changing.

One cannot speak geographically without speaking of places, and one cannot speak effectively of places without considering the regional concept and the pattern of areal relationships. Special analysis of industrial resources, transportation, agricultural types, settlements, or political factors is not only incomplete but liable to serious error until it can be applied within a variety of regional frameworks and regional contexts. This means more than superimposing data or ideas on a landscape, or extracting them from it; the regional whole must be understood. Regional models must be constructed against which any assertion or study or discovery with regional relevance can be measured.

The progress of knowledge is not served by insulating the boundaries of convenience which separate its various branches. No student of society is self-sufficient, and all of the branches of study are interdependent. Within geography, any sound analysis must make thorough use of both the topical and the regional approaches as well as depending on contributions from other disciplines. But it is the nature of geography that at the heart of all its work is the concept of the region.

Questions for Further Study and Discussion

1. Find several other examples of regional cores. How do they differ from each other? Can you find regions, or political states, without clearly defined cores? With two or more cores of approximately equal prominence? With one dominant core and other subsidiary ones? What conclusions do you draw from these examples?

2. Divide the United States into regions at three different orders of magnitude. Characterize and distinguish each region, draw the necessary lines around it, and define the regional core of each. To what extent do you find it appropriate to use state boundaries? Why are state boundaries not more generally appropriate? How do you account for the instances where state boundaries do seem to coincide reasonably well with regional breaks? In the regional divisions you make, what kinds of problems arise over transition zones or zones of indifference?

3. Apply a three-level regional division to your own state.

4. What characteristics or factors, and in what order of importance, do you find it necessary or useful to consider in making the regional distinctions in the preceding two questions?

5. What kinds of single-factor or functional regions and how many would you distinguish in the United States?

6. How would you describe and delineate the core area or areas of the United States? To what extent are there subsidiary or evenly balanced separate cores? How would you apply the same analysis to Germany, and to Japan? Why in Germany and Japan is a single core less dominant than in Britain or France?

7. Select and discuss two examples of settlements whose present status must be explained primarily in terms of historical or dynamic rather than static factors.

8. To what extent is it reasonable to regard geography as "human ecology"? Why is or why isn't such a label appropriate?

Selected Samples for Further Reading

Anderson, M. *Splendor of Earth.* London, 1954. An anthology of literary passages which describe landscapes, many of them vivid and conveying a sense of reality which is an important part of geographic study.

Berry, B. "Approaches to Regional Analysis," *Annals of the Association of American Geographers,* LIV (1964), 2–11. A compact argument for examining geographical data as keys to broader "systems," and for regional analysis based on such an approach.

Broek, J. O. M. "National Character in the Perspective of Cultural Geography," *Annals of the American Academy of Political and Social Science,* CCCLXX (1967), 8–15. The quality of regions often looked for by other social scientists, as seen by a geographer.

Campbell, R. D. "Personality as an Element of Regional Geography," *Annals of the Association of American Geographers,* LVIII (1968), 748–59. A useful effort to apply some recent ideas in psychology to the study of regional environmental effects on attitudes and behavior.

Darby, H. C. "The Problem of Geographical Description," *Transactions and Papers of the Institute of British Geographers,* 1962, 1–14. The demanding job of adequate regional description, and some suggestions from a leading British practitioner.

Dickinson, R. E. *City, Region, and Regionalism: A Geographical Contribution to Human Ecology.* London, 1947. Primarily concerned with the city and its regional manifestations, but dealing also with wider notions of regional form.

Grigg, D. "The Logic of Regional Systems," *Annals of the Association of American Geographers,* LV (1965), 465-91.

Heslinga, M. W. *The Irish Border as a Cultural Divide.* New York, 1963. A study of the regionalism and regional feeling which underlie this boundary.

Highsmith, R., ed. *Case Studies in World Geography: Occupance and Economy Types.* Englewood Cliffs, N. J., 1961. Sample studies of small areas from different parts of the world, as illustrations of larger relationships.

James, P. E. "Toward a Further Understanding of the Regional Concept," *Annals of the Association of American Geographers,* XLII (1952), 195–222. Useful reflections by a leading practitioner of the regional method.

Jensen, M., ed. *Regionalism in America.* Madison, Wis., 1951. A collection of articles and essays, mainly by sociologists, on various aspects of American regionalism and American regions.

Minshull, R. *Regional Geography.* London, 1967. A textbook treatment of the variety of meanings associated with *region, regionalism,* and the *regional method,* with illustrations and discussion of different sorts of actual and theoretical regions.

Odum, H. W. *Southern Regions of the United States.* Chapel Hill, N. C., 1936. An attempt at holistic analysis of perhaps the most distinctive human region in America.

Pounds, N. J. G., and Ball, S. S. "Core Areas and the Development of the European State System," *Annals of the Association of American Geographers,* LIV (1964), 24–40. Core areas primarily as historical growths in each of the major European states and their role in the creation of a cohesive national region.

Prescott, J. R. V. "The Function and Methods of Electoral Geography," *Annals of the Association of American Geographers,* XLIX (1959), 296–304. Voting patterns as indicators of both political and nonpolitical aspects of regions and of their delimitation, especially in areas whose attachment to a larger region or to a political state may be ambiguous or ambivalent.

———. *The Geography of Frontiers and Boundaries.* Chicago, 1965. A good general treatment of concepts and problems, with a series of case studies.

Russett, B. M. *International Regions and the International System: A Study in Political Ecology.* Chicago, 1967. A good statement of the regional concept and a series of imaginative applications to national political units—despite its misuse of the word *ecology.*

Spate, O. H. K. "Region as a Term of Art," *Orbis,* I: (3), 1957. A brief and witty, often biting, essay on the various uses, misuses, and meanings of the term *region.*

Ullman, E. L. "Human Geography and Area Research," *Annals of the Association of American Geographers,* XLIII (1953), 54–66. Geography's role in the study of regions, varied approaches to that problem, and the need for synthesis.

———. "Regional Development and the Geography of Concentration." In J. Friedman, and W. Alonso, eds. *Regional Development and Planning,* Cambridge, Mass., 1964, pp. 153–72.

Whittlesey, D. S. "The Regional Concept and the Regional Method." In P. James et al., *American Geography, Inventory and Prospect.* 2nd ed., Syracuse, 1964. The best summary statement, and a further bibliography.

7 The Climatic Influence

IT IS DIFFICULT to consider adequately any aspect of man's settlement on or use of the earth without considering the environmental influence. But it is perhaps best to think of the environment as a *permissive* rather than as a causative factor, in terms of the range of choice or the kind and degree of opportunity which it offers to man. In a harsh setting like the polar ice cap, man has little choice. Without a highly developed technology imported from some other area, he can maintain himself only in a strictly circumscribed manner, and there is little opportunity for him to specialize or to branch out into other activities. In harsh environments human settlement is usually characterized by a low standard of living and a primitive level of development. The business of staying alive is demanding enough to leave no room for the accumulation of a surplus or for the development which depends on a surplus.

THE RANGE OF CHOICE

In a milder physical situation the environment offers a wider range of choice. The technically, economically, and culturally most developed societies have arisen in environments offering such a wide range. Surpluses, specialization, and exchange have been possible. But even so, physical limits operate, and the environment may help to make some choices more profitable or more successful than others. One cannot grow cotton or rubber profitably in northwest Europe, for all of Europe's economic development. Northwest Europe is also for the most part less well suited by comparative advantage than other areas for bulk agriculture production of the major cereals, despite the fact that Europe is typical of the mildest of all environmental situations, the ones with the greatest range of choice. The role of environment in influencing the human development which takes place in such permissive areas can be understood, like so many other things, only in connection with spatial relations. Specialization is possible only if there is cheap access to other areas. Europe now feeds itself in large part from the agricultural surpluses of Australia, Argentina, North America, and the tropics, carried to Europe by cheap sea transport developed during the past century. Northwest Europe has thus been able to concentrate on the specialties in which environment has helped to give it a comparative advantage, most importantly trade, manufacturing, and high-value agriculture for the huge local market. Wheat yields better in the climate of northwest Europe than in the climate of Dakota or Australia, but Europe's location and industrial resources make trade and manufacturing even more profitable than wheat growing, whereas in Dakota or Australia wheat has few rivals even though it is less productive than in Europe.

In the discussion of Switzerland it was pointed out that in a dramatic environment like the Alps one expects to find a clearer causal relationship between physical factors and human development than in milder situations where there is more scope for many other factors. Certainly in the rest of northwest Europe, or in northeast United States, physical conditions are much less important as controls for the whole economy and society than in Switzerland, and much less apparent; they are more permissive. Physical conditions are often involved in smaller-scale matters, such as the site of a city, the location of market-garden areas around it, or the sparse population of rugged areas. But environment is necessarily involved also in the relations between places, making access easy or difficult and contributing importantly to regional differences and to specific economic or cultural systems. The nature of land use, which environment helps to determine, and the total resource complex may stimulate or minimize complementary exchange between regions. North and south Europe have a complementary exchange relationship, illustrated by the exchange of Italian lemons for German coal, in large part because of differences in their physical environments. Environment operates in these ways in every part of the world, but where the range of choice is greater, direct environmental influences are less strong. In nearly every situation involving man's distribution on and use of the earth, however, and in most aspects of his culture, the physical environment and man's interrelationship with it have had some part to play, either directly or as part of spatial interaction. This is by no means the same as saying that one can explain everything about mankind in these terms, but only that the physical environment is usually one of many relevant factors.

HAS MAN CONQUERED NATURE?

It is possible to argue that man becomes more rather than less dependent on the physical environment as his technology, economy, and society become more highly developed. Although he may be able to irrigate the desert, heat his house, fertilize the soil, chop down the forests, and tunnel under the mountains, he does not remove the limiting conditions which these phenomena imposed upon him but merely exchanges one set of conditions for another. It costs money to "conquer the forces of nature," usually a great deal of money. It pays to spend this much money only under certain circumstances, part of which are usually a reflection of the physical environment. Tunnels pay only with an existing or potential high volume of traffic, irrigation pays only for a present or potential highly productive agricultural system, and so on. Each of these is in turn dependent on what the physical base and its relations in space will allow. Once the technical improvement has been made, man's choice is also limited. He must maximize the return from such a dearly bought piece of property, and he can therefore use it only for certain restricted high-value or highly productive goods or services. Each such technological alteration of the environment usually supposes or requires a more or less complex system of economic, social, and political organization to create the alteration, to maintain it, and to operate the production system dependent on it. For example, in many of the irrigated areas of monsoon Asia water control required and helped to perpetuate a powerful state bureaucracy, and in most Asian societies the individual entrepreneur was in a weak position. Changes which man's technology enable him to make in the environment may also create new resources which fundamentally affect human society. On the other hand, man may destroy resources or alter the environment in such a way that human activities suffer or are limited.

Before man had these technical abilities, he was, of course, obliged to walk around or over the mountain or avoid the desert, unless he chose to be a nomad. But the mountain and the desert really meant less to him, in terms of effort or of dollars or of the distribution of routes and settlements or of economic production, than they do now. As specialization proceeds, interregional exchange also becomes increasingly important. Each region is concerned with the economic, political, and cultural complex of many other regions, some of which may be remote but complementary. Each region is thereby concerned, too, with the environmental basis of other regions and with the aids or barriers to movement across the intervening space. The weather in the Australian wheat belt is important in the markets of Liverpool and Chicago and on the farms of Dakota; the fact that these scattered places are connected by the cheap volume transport created by the industrial revolution[1] is responsible for their mutual sensitivity to each other's environment. The transformation of tropical rainforests into rubber plantations has made the local environment far more important to man than before the forest was altered and has also made Detroit sensitive to what happens in Malaya or Ceylon.

Specialization also means that man is dependent on many more resources than he used to be. To the American Indians of the fifteenth century, iron ore in Minnesota or coal in Pennsylvania meant nothing. They had no worries about the remoteness of the ore, the cost of its shipment to population centers, the freezing of the Great Lakes, or the cost of mining and shipping coal from the Pennsylvania mountains. If

---

[1]See Chapter 13 for a discussion and definition of "industrial" and "industrial revolution."

their wells or streams ran dry, they could always move and did so frequently as a matter of course. A steel mill is much more tightly bound to its site and to the physical conditions existing there to which it is extremely sensitive, especially the supply of tremendous quantities of fresh water for cooling and flushing purposes.

Each economic specialization which man develops makes him more sensitive to both local and distant environments, because it increases his dependence on specific or varied resources and his dependence on exchange. A modern fruit orchard, for instance, contrasts markedly with the Indian cornfield or hunting ground which might have occupied the same site three hundred years ago. The fruit farmer must be concerned with many local conditions which were more nearly matters of indifference to the Indian—early frosts, air drainage, insects, hailstorms, and other things in which the environment is partly reflected, such as the price of land and the availability of low-cost labor at picking time. He is also concerned with the same conditions in other fruit-growing areas with which he competes in the national market. The price of New York State apples (and hence the fruit grower's profit) will depend in large part on the quantity and quality of the Washington State apple harvest and the price tag attached to it; it will also depend on the condition of overseas markets and on the competition from other fruits, or simply from other food items, for a share of the housewife's food budget. Each of these will in differing degrees reflect environmental factors.

There is, of course, one vital respect in which man has gained the upper hand over his environment. He may still be limited by the environment in his economic decisions and developments, but by and large he no longer starves to death because nature has been niggardly or vacillating. Exchange evens out the regional and temporal humps and hollows in production at the same time that it makes man more sensitive to the environment. Perhaps this is progress enough to say that man has conquered nature, but it is a limited conquest which has done the reverse of eliminating the environment as a factor in man's affairs.

Each culture and each period of human development may be regarded as a different type or a different stage of adjustment to a given environment. Although man may make this adjustment more productively as his technology improves, the need for adjustment is not thereby disposed of and may indeed become more demanding as a greater variety of environmental factors or resources become important. It should also be clear that the process of regional specialization which technological improvements and cheap exchange make possible continues to create greater and greater regional economic diversity over the world, at least in terms of production, rather than making regions more alike. This is so despite the degree of uniformity which improvements in technology and exchange may also spread in many other respects. We are no longer all farmers, nor do all agricultural regions produce the same kinds of things or attempt to produce everything they need. Though industrial cities may soon be world-wide in their distribution, and with them the other trappings of an industrial-commercial civilization, each larger region and each subregion within it grows more rather than less distinctive as it becomes more specialized or as it is increasingly able to make special use of its total set of environmental and other advantages. Technological change thus leads in this sense to an increase rather than a lessening of environmental influences on man, as it also leads to an increase in spatial interaction, which is in part an expression of influences between different environmental bases. Location is important only if there is interaction, if there are other places related to the place in question. As these relationships increase, so do the influence of location and the influence of varied environments increase in their impact on man's activities.

THE ENVIRONMENTAL COMPLEX

There are five basic elements of the physical environment which tend to operate as a unit: climate, vegetation, soils, landforms, and water resources. Usually in any given area they tend to be consistent with each other. A certain climate will tend to be accompanied by a characteristic set of vegetation, soils, landforms, and water resources. Mineral resources other than water are not related to these five in their distribution or are related in less current or less obvious ways. They do not occur consistently or directly as part of the same physical complex. Current climate, for example, has no consistent relationship to coal deposits. Ores and other mineral deposits are certainly part of the physical environment, but whether they are usable resources in a particular case depends as much on economic, locational, and cultural considerations as on the physical facts. Many mineral deposits which are loosely called "resources" are not so at all, because they are not economically usable; the word *resources* could be better applied to climate or soils or to human skills. The meaning of resources will be discussed in detail in Chapter 13, which also includes a section on the biotic resources—animals, fish, insects, birds, and plants—as an important part of the physical environment. For the present the discussion will consider the physical environment in terms of the five elements of climate, vegetation, soils, landforms, and water resources.

THE MAJOR ROLE OF CLIMATE

Climate is by far the most important aspect of the environment; most of the other environmental factors are in part reflections of it. Few human developments can ignore climate or be independent of it, and climate's permissive or conditioning influences are apparent far beyond the obvious sphere of agriculture. Climate more than anything else about the environment sets the limits of what mankind can physically or economically do. It offers or denies him opportunity or special advantages and is therefore directly involved in the distribution of population over the earth and the particular economies and societies which have arisen in different regions. The great empty spaces of the world, from a human point of view, are predominantly the work of a harsh climate, and the great clusters of population are generally coincident with areas where the climate is broadly permissive. Within the range of choice which a permissive type of climate offers, man may develop much or little, in accordance with a number of non-climatic factors, some of them pertaining to other aspects of the local environment, some to relations in space, and some to matters such as cultural tradition, political management, or social structure. But climate may be said to set the stage on which man may play his role by the manipulation of other less uncompromising factors.

Man's Control over Climate

It is still reasonable to assume that man can do very little to change climate. He can heat his house or install air conditioning, carry an umbrella or wear warm clothing and thus be saved some physical discomfort. But he does not thereby escape the economic limitations which climate imposes on what he can profitably do for a living; rather he compounds limitations by spending money to shield himself from the climate and thus makes it necessary to produce that much more from the area where he lives and works. He can irrigate or he can install orchard heaters, but this merely substitutes economic limits for physical limits, with climate remaining at the root of both. For the most part he must accept climate as something over which he has little or no broad control and to which man must fit himself.

There do not seem at present to be any strong possibilities of technical developments which would in the foreseeable future enable man to control or modify climate except on the restricted scale suggested by the above examples. Experiments with artificial rain making by seeding or bombarding clouds from an airplane have so far not convinced most meteorologists that this is even potentially a technique which could alter climate (as opposed to daily weather) on a significant scale over a large area. The wind systems of the upper atmosphere, on which the broad mechanics of weather and climate appear in large part to rest, are massive forces which seem difficult if not impossible to control and at present not even easy to delineate in detail. The force or energy represented by a single hurricane or even by a normal rainstorm is equal to many thousands of nuclear bombs. It is conceivable that a strategically placed and timed nuclear bomb might be used to "trigger" a desired change in weather by upsetting the local atmospheric balance on a small scale. One may expect increasing accuracy of forecasting on the basis of statistically accumulated observations and especially with the help of new weather satellites, which can improve long-range forecasting. This can be of great importance, not only to farmers but for the better periodic planning of the economy as a whole. But man does not yet begin to know enough about the atmosphere and its pattern of forces to attempt effective manipulation.

CLIMATIC CHANGE AND FLUCTUATION

Climate is not constant. It moves through a hierarchy of cycles, from the obvious change between day and night through the changes of the seasons (weather is simply daily climate) to longer cycles of years, centuries, and geologic periods. Between the great ice ages and the daily or seasonal changes in climate there is a gradually decreasing scale of change until one reaches a point somewhere near the beginning of the Christian era, when it can probably be said that the climate of the world as a whole became substantially the same as it is today. Fluctuations in climate on a world scale since that time have been minor compared with what went before, and such fluctuations have probably also returned periodically to a relatively constant mean. On a graph, temperature, for example, might be represented schematically as it is on Figure 11. In time the curve may drop significantly again as the earth enters a new glacial period, or it may rise significantly. Such time measurements would take the discussion not only beyond man's present technical means of understanding such things but also beyond the span of time with which this book is concerned. The earth does not seem to be faced with an ice age in terms of the next thousand years, and it is far enough removed in time from the last one so that the geographer needs to be concerned with it only when dealing with early man or early civilizations. This is so even though the earth may now be in what is called an interglacial period or on the threshold of a major rise in temperature or change in rainfall which would destroy or distort the average of the last two thousand years. Larger questions of climatic change in terms of thousands of years are extremely interesting, but the observable time span of what may be called human civilization is smaller. Climatic changes in the course of the last glaciation and its retreat were of transcendent importance for man as he then existed and as he developed (see the item by Carl Sauer in the reading list at the end of Chapter 8). It is a suggestive idea that *Homo sapiens* owes some of the flexibility and adaptability which have enabled him to multiply and spread over the world and to use the physical environment so effectively to the fact that early in its development the human species was obliged to adjust to drastic changes in climate, and that it did so successfully where many other species failed. But for most of recorded human history it can be assumed that despite

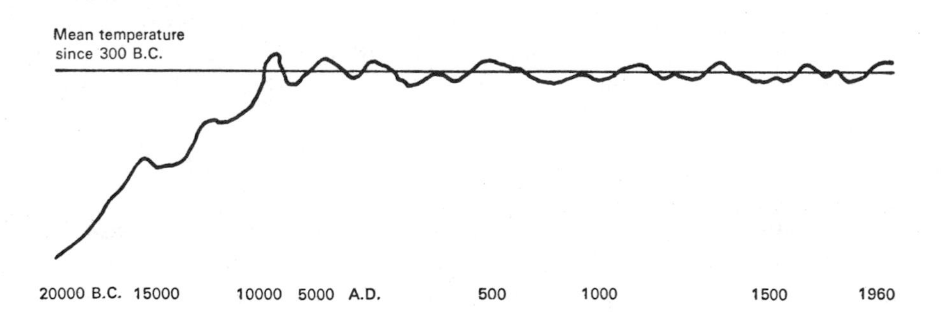

Figure 11. World Temperature Trends.

continual fluctuations, the climate of the world as a whole has been broadly consistent.

Climatic fluctuations have obviously been important to man, even though in the long run fluctuations have tended to balance each other out during the last two or three thousand years. Especially in areas whose normal climate is marginal—semi-arid or semi-boreal (cold)—a slight change in rainfall or temperature can be crucial, enough to damage agriculture, for example, or temporarily to remove some of the basis for human settlement. Such fluctuations have attracted the attention of many people who have used them as evidence of climatic change, which they then apply to explain changes in human society. But it is the nature of climate to fluctuate. Progressive and consistent change is another matter. The line between fluctuation and progressive change is difficult to draw; it depends on the time interval used. The change from winter to summer is progressive in a sense, and cold years may be followed by warm years. But over a century or two climate is more likely to fluctuate around a relatively constant average. During the past two thousand years, alternating series of wet and dry or cold and warm years, decades, and centuries have probably not represented enough progressive change, on the basis of the scattered and unsatisfactory evidence which is available, to alter fundamentally the climate of most areas, or to provide a satisfactory basis for explaining most of the changes in human settlement during that period. There are very few weather or climatic records available before about A.D. 1750; earlier records which have been preserved are not easy to interpret accurately because the standards of observation or measurement are seldom clear or consistent. Geologic evidence is largely useless on a time scale of centuries, and even the evidence of changes in fauna and flora is not easy to read with precision. Other factors than climate have usually been far more important in explaining human changes.

CLIMATE AND HUMAN DEVELOPMENT

Climatic determinism has also been used repeatedly to explain current human differences over the world. Climate has been more abused than most of the environmental elements in this respect, because its force is often so important and often so clear that it tempts the investigator into simple, single-factor explanations. The Greeks, including Aristotle, Herodotus, and others, wrote of the determining effects of climate on the nature of political organization, individual temperament, morals, and tastes. Aristotle stated that the same latitudes anywhere in the world produced the same things, and that the lowest latitudes were the most richly productive, especially in precious metals. This was one of the reasons why Columbus followed a southerly route from Spain to the New World. Within the Mediterranean Basin or near it the Greeks were familiar with a variety of climates. Climate was the most obvious and to them the most important difference between the various areas which they knew, and they ascribed to climate the equally great human differences which they observed. Climatic determinism has been a persistent idea through the Middle Ages in Europe and into the modern period, when people still speak poetically of "climes" when they really mean places or areas or cultures. Several modern writers have attempted to explain differential human development and even areal differences in human temperament and abilities in terms of climate. It is suggested, for instance, that the greatest economic and intellectual development and the home of "progress" is in the "temperate" lands like northwest Europe and northern United States. It is also explicitly stated that in less "temperate" climates there has not only been less development, but human energies and mental powers, dictated by climate, do not allow it.

This would be a laughable thesis if it had not been taken so seriously by so many people. It has the appeal not only of a simple solution to a complex problem, but of a not very subtle flattery of those who happen to live in what they call the "temperate" lands and an implied insult to everybody else. It is certainly a blatant example of egoism or ethnocentrism; at best it applies to all areas the provincial standards of one. It may be true that the European, for example, is sluggish in India or in China, but this does not mean that the Indian or the Chinese is sluggish, and no objective measures could deny that both of these areas produced civilizations which were in most respects superior to Europe's outside the Mediterranean until about two hundred years ago. From the Asian point of view, or seen from the genuinely temperate tropics, Europe and North America are neither temperate climatically, nor, until the last two or three centuries, particularly progressive. Few people have suggested that climatic change is responsible for the recent relative decline of the East, but there have been suggestions that north Europe's relative backwardness until the modern period was the result of a colder and wetter climate which has since improved.

Evidence for this, in terms of progressive and consistent change, is slim. But more important, there have been other factors at work which adequately account for the change in Europe: notably the industrial revolution since the eighteenth century, whose occurrence was recent and was certainly not due to climatic change; and the earlier Age of the Discoveries, which brought about the readjustments discussed in Chapter 2. In contemporary terms, Europe's "progress" is said to result from the stimulating effects of its climate on human inventiveness and drive. But this is clearly reasoning after the event of the discovery and use of industrial resources in Europe and of the advantages of location which became operative after the Renaissance.

If the climate were such a basic determinant, how is one to account for the fact that California was inhabited by one of the most primitive of all the American Indian groups? That hot, sticky Canton, where frosts almost never occur, was and is economically, culturally, and technically more highly developed than Yunnan with its bracing climate? That northeast United States, with a climate similar to north Europe's, was inhabited by stone-age savages until the seventeenth century A.D., and the emperor Akbar's technically sophisticated India and his magnificent capital at Fatehpur Sikkri on the baking plains of Uttar Pradesh were nearly contemporary with the meager first Thanksgiving in bleak New England? That just as brilliant modern technicians, scientists, poets, scholars, and statesmen can be found in tropical countries as in the "northern temperate lands"? It is true that different people and racial groups have different tolerances for heat and cold, different dietary requirements, metabolisms, blood pressures, and physical sizes, and also that anyone is less vigorous mentally or physically if he is infested with intestinal parasites or malaria or lives on an inadequate diet. While these things are clearly related to climate, diseases or dietary inadequacies can be removed or greatly reduced, as they have been in the last two hundred years in the West. There are no substantial physiological grounds for assuming that an Indian, for example, with proper diet and health, is not or cannot be as physically and mentally vigorous in his climate as the European or American in his.

There are good reasons for the present gap in economic development between northwest Europe or North America and the tropics of Asia or South America. Climate has something to do with this gap, not because it directly limits people's minds or physical capacities but because under certain circumstances it limits economic opportunity. These limiting effects can be and are being lessened by technical development. While it seems unlikely that India, for example, will ever in the foreseeable future equal or surpass the income levels or per capita economic productivity of North America or northwest Europe, the reasons for this are only partly climatic. More importantly, they are a reflection of other resources, including location, and of a huge and still growing population.

Climate cannot be used to explain everything about regional differences, nor can it safely be regarded as a direct or consistent influence on human capacities by any universal measure of accomplishment. Such a mechanistic type of analysis is intrinsically to be distrusted and is easily disproved. But this is not to deny the importance of climate. The climatic influence is less direct or perhaps less personal. But indirectly, in terms of the economic opportunity or special advantages which climate may offer or deny, it is a basic factor.

8
The Workings of Climate

THERE ARE TWO PRINCIPAL elements of climate[1]: energy (usually measured in terms of temperature) and moisture (including water vapor, solid and liquid water, and the processes of condensation, precipitation, and evaporation). Precipitation may occur as mist, sleet, snow, or hail, as well as rain, but evaporation is equally important. Annual averages are the most commonly used measures for each of these elements. But the seasonal distribution of temperature and precipitation, the nature of the precipitation, and the variability in temperature and precipitation from year to year may be even more important. Obviously, an area which gets most of its rainfall in winter, or in only one season, is less well off than one whose rainfall is more evenly distributed throughout the year or which has most rainfall during the growing season. The nature of the precipitation is also important. A slow gentle shower, most of which can be absorbed by the soil, is much more desirable than a violent downpour, which erodes the soil, may damage crops, and may run off the surface into stream channels before it has a chance to soak into the ground. An area with twenty inches of gentle rain a year may be much better watered than one with forty inches of violent showers.

Annual variability, especially in precipitation, is usually proportionately highest in areas where its total average amount is least. In the marginal areas where rainfall is most crucial, average annual variability may be as high as 30 to 40 per cent, as it is for instance in much of north China. With an average annual rainfall of twenty inches, thirty or more may fall in one year (which may wash away the crops or cause floods), and ten or less in the next year. Where the annual averages are higher, proportional variability is usually much less but may

---

[1]Atmospheric pressure and winds are also important elements of climate, but they will be treated as they arise in the course of the following discussion.

still be a major problem. Plants, animals, and men adjust to long-term averages of climate. They may be seriously upset by shorter-term fluctuations, or where such fluctuations are great and repeated, plant, animal, and human life must adjust accordingly and be limited accordingly. Variation in seasonal incidence can also be critical. An unseasonably cold or warm spring has widespread and usually harmful effects on all forms of life. Summer rains which on an average come early in the growing season may arrive too late or may be harmful or useless because they come too early. The extremes of climate may thus be more important than the averages.

Temperature is important mainly for its extremes. The length of the frost-free period, between the last killing frost in spring and the first in autumn (which of course varies from year to year, with the greatest variability toward the marginal or more limited areas, as with precipitation), is probably the most important measure of the effects of temperature. The growing season is a little shorter, since for most plants the minimum vegetative temperature is several degrees above 32° Fahrenheit. Temperatures during the growing season are also important, and the amount of energy thus available is directly related to plant growth. Areas free from frost at all seasons have a great vegetational, agricultural, and general economic advantage; the longer the frost-free period, the less the degree of frost, and the greater amount of energy in heat terms, and of sunshine, other things being equal, the greater the advantage.[2] Extremes of heat and freezing cold of course affect many things outside agriculture, including transportation, housing, and manufacturing costs.

Temperatures also have a direct bearing on the effectiveness of precipitation, because they help to determine evaporation rates and the degree of atmospheric moisture, or humidity. Low humidity may have serious desiccating effects, and high humidity may prevent desiccation, quite apart from actual precipitation. In a hot place like India, for example, twenty inches of rainfall per year may still leave an area semi-arid, whereas parts of England which are obviously well watered receive about the same amount. The difference in effectiveness in this case is also accounted for in part by the fact that India's climate is monsoonal and that the rainfall is therefore heavily concentrated in one season (summer). Generally, the lower the temperatures (aside from seasonal variability), the less rainfall is needed to produce a given amount of moisture in the soil. Mark Twain once defined a cool day in India as one on which "the door knobs do not actually melt, but merely become mushy"; he might have added that on such days the rain does not actually turn to steam as it falls but waits until it strikes the ground.

---

[2] Heat may of course be excessive and sunshine may be restricted by humidity or cloud cover, or by day length. Much of the tropics may thus be less productive agriculturally than areas where temperature and moisture are in better balance and summer days are longer. The effective growing season is a result of a complex of factors. For a useful discussion, see Jen-hu Chang, "A Critique of the Concept of Growing Season," *The Professional Geographer*, XXIII (1971), 337-40.

PRECIPITATION

Rainfall, or precipitation in general, is the more important of these two elements of climate, though it might be hard to choose between the Sahara Desert and the poles as rival samples of what each can produce at the extremes. Precipitation can occur only through the cooling of an air mass and in no other way. As an air mass is cooled, it loses some of its capacity to retain moisture, which is precipitated out of it by condensation. The amount of water vapor which an air mass can hold depends on the temperature of the air mass and increases as the temperature rises. There are three principal ways in which an air mass is cooled, and one type of rainfall is associated with each (see Figures 12, 13 and 14). Any one storm may involve all three of these processes, although they are often distinct.

1. *Cyclonic or Frontal.* The gradual convergence and contact of a warm air mass and a cold air mass along an extended front may result in the cooling of the warm air mass in one of two ways. (*a*) The cold mass, being denser, slides under the warm mass, which is thereby forced to rise over it, usually producing heavy showers. (*b*) The warm mass overtakes a cold mass and slowly rises over it; rainfall produced in this way usually falls as slow, gentle showers, often extended through one to three days. Cyclonic or frontal rainfall is characteristic of the middle latitudes, including most of North America and Europe, where warm and cold air masses from south and north respectively converge in low-pressure cells which move slowly across the land from west to east. Frontal showers may, however, occur in other latitudes, wherever warm and cold air masses meet, and in any latitude the convergence of air masses whose temperatures differ markedly may also be accompanied by relatively violent showers and by turbulence and thunder.

2. *Convectional.* The rapid rising of air which has been heated by direct or indirect contact with the surface of the ground may cause the equally rapid precipitation of the moisture which it contains. Hot air rises, and when an already warm body of air which may therefore hold much moisture is further heated quickly, it soon rises high enough to condense most of its moisture at once, which falls as a violent shower. This is the basis of thunderstorms, which are generally limited to places or to seasons which are hot. In large areas of the tropics they are the predominant source of rainfall, and in colder latitudes they are concentrated in summer or are absent.

3. *Orographic.* The rising of an air mass as it strikes a topographic barrier may result in precipitation. Air cools as it rises because of the decrease in atmospheric pressure as altitude increases and because the surrounding upper air is usually cooler. (The greater the pressure on any gas, the higher its temperature, other things being equal; cooling produces precipitation.) There is usually heavy rainfall on the windward slopes of mountain ranges if they lie in the path of moisture-bearing winds and a much drier area on the leeward side, often a desert. In the United States the best example of this phenomenon is provided by

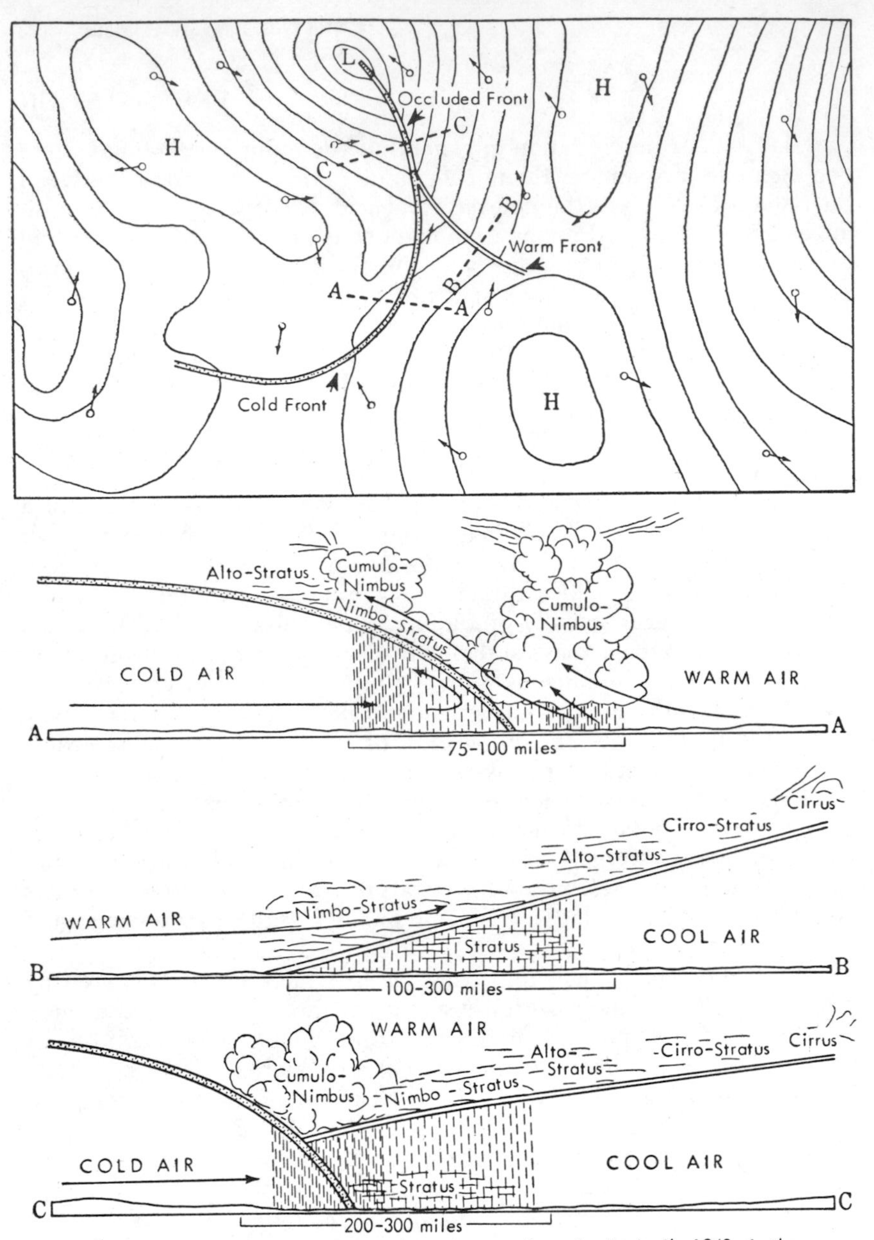

This diagram was prepared from an actual weather map for a day in April, 1942, in the north central part of the middlewestern United States. High pressure areas of relatively colder air are indicated by H, low pressure area of relatively warmer air by L. Winds, or exchanges of air, blow from high to low pressure areas. Their velocity depends on the pressure gradient, which is the result of the difference in pressure plus the distance between two air masses of different atmospheric pressure. The fronts indicated on the weather map and in the three cross-section diagrams are discontinuity zones where cold and warm air masses meet. Sharp changes of temperature, moisture content, and atmospheric pressure occur along these fronts; rapid shifts in wind direction and velocity are also characteristic along fronts.

Figure 12. Cyclonic Precipitation.

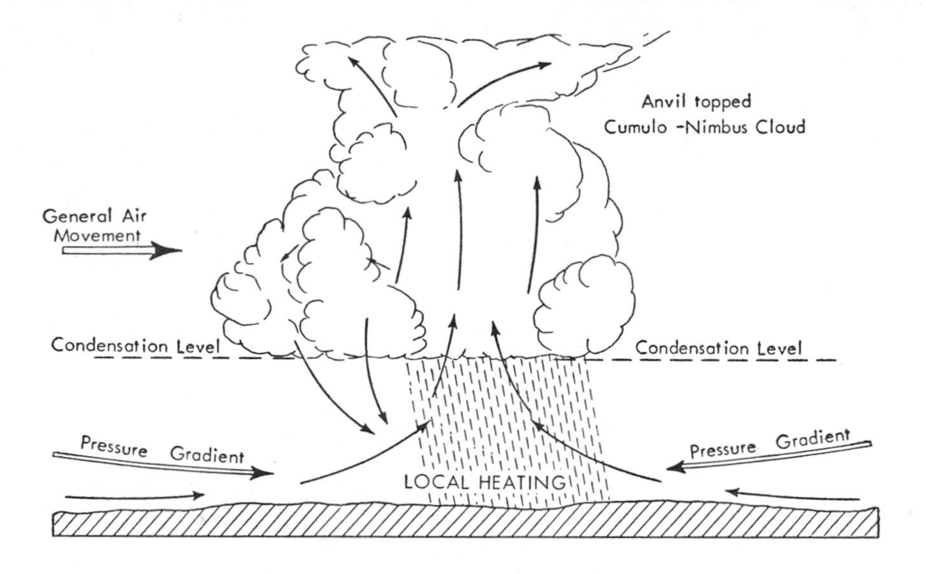

Figure 13. Convectional Precipitation. Air temperature, at and above the condensation level, determines whether precipitation is in the form of water (rain) or snow.

the coastal ranges of Washington, Oregon, and California; on their seaward slopes annual rainfall is generally above sixty inches and in several places exceeds a hundred inches; the landward side lies in what is called a *rain shadow,* and annual rainfall is under twenty inches. In large areas it is less than ten, a semi-arid or desert country. The mountains have drawn a sharp climatic line which is readily apparent in traveling across the Cascades or the Sierras. Air masses, which in this latitude are moving mainly from west to east, are milked of their moisture by the forced rise over the mountains. As they descend on the leeward side, they are warmed in the descent by rising atmospheric pressure and may often absorb more moisture than they release. This is the origin of the *foehn* or *chinook* winds, which often melt snow and cause avalanches or may parch a landscape. Similar situations, including rain shadows, are to be found wherever mountains lie across the path of the prevailing moist winds.

CLIMATIC TYPES

Climates may schematically be divided according to latitude, largely in accordance with the amount and directness of the sunshine which each area receives. Sunshine, or insolation, varies in amount and in intensity according to latitude because of the inclination of the earth on its axis, as shown in Figure 15. Climates may also be classified as maritime or continental on the basis of their relationship to the sea.

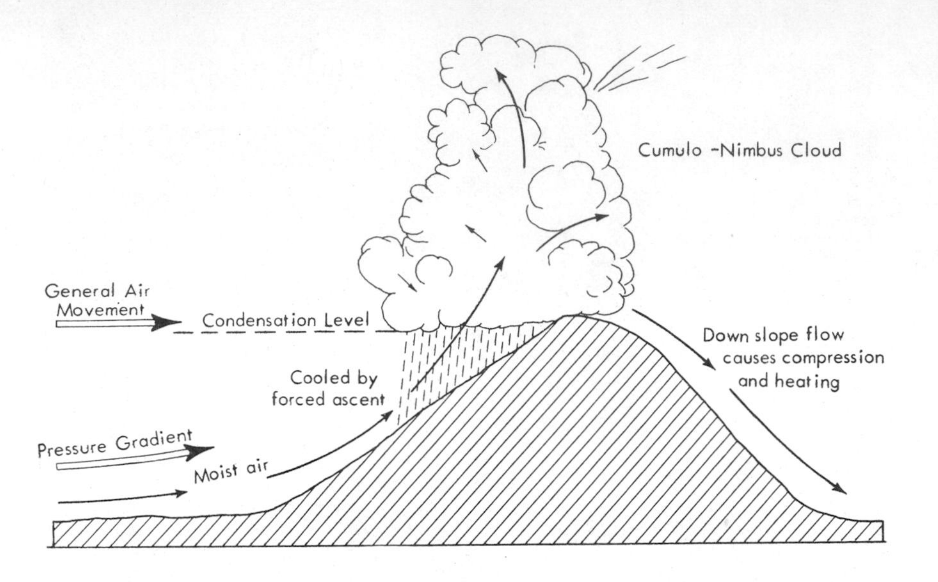

Figure 14. Orographic Precipitation. Air temperature, at and above the condensation level, determines whether precipitation is in the form of water (rain) or snow.

Briefly stated, similar climates tend to occur in similar latitudinal and continental positions. For example, northwest United States and northwest Europe have mild moist climates which are roughly the same. Both are at similar latitudes and both are strongly influenced by the sea; both are examples of maritime climates. Central Asia and Kansas are dry or semi-arid and have cold winters and hot summers; both are at similar latitudes far from the sea, and both are examples of continental climates. Latitude and continental position are far from being the only factors affecting climate, although they are probably the most important. Any division or classification based only on these two measures is grossly oversimplified. Maps of world climates use a classification system based on many different factors, which are clearly indicated. With such a system and map to refer to periodically, this discussion can deal in broader terms which, to begin with, may be clearer. The effect of latitude on climate is clear enough, and most readers will have had personal experience with it, within North America or elsewhere. The effects of the sea, which are responsible for the difference between maritime and continental climates and for many other climatic aspects, require some discussion.

Climatic Effects of the Sea

The sea is a great modifier of temperatures, as well as being the source of most rainfall and atmospheric moisture. Water heats and cools much more slowly than land, partly because it circulates, whereas

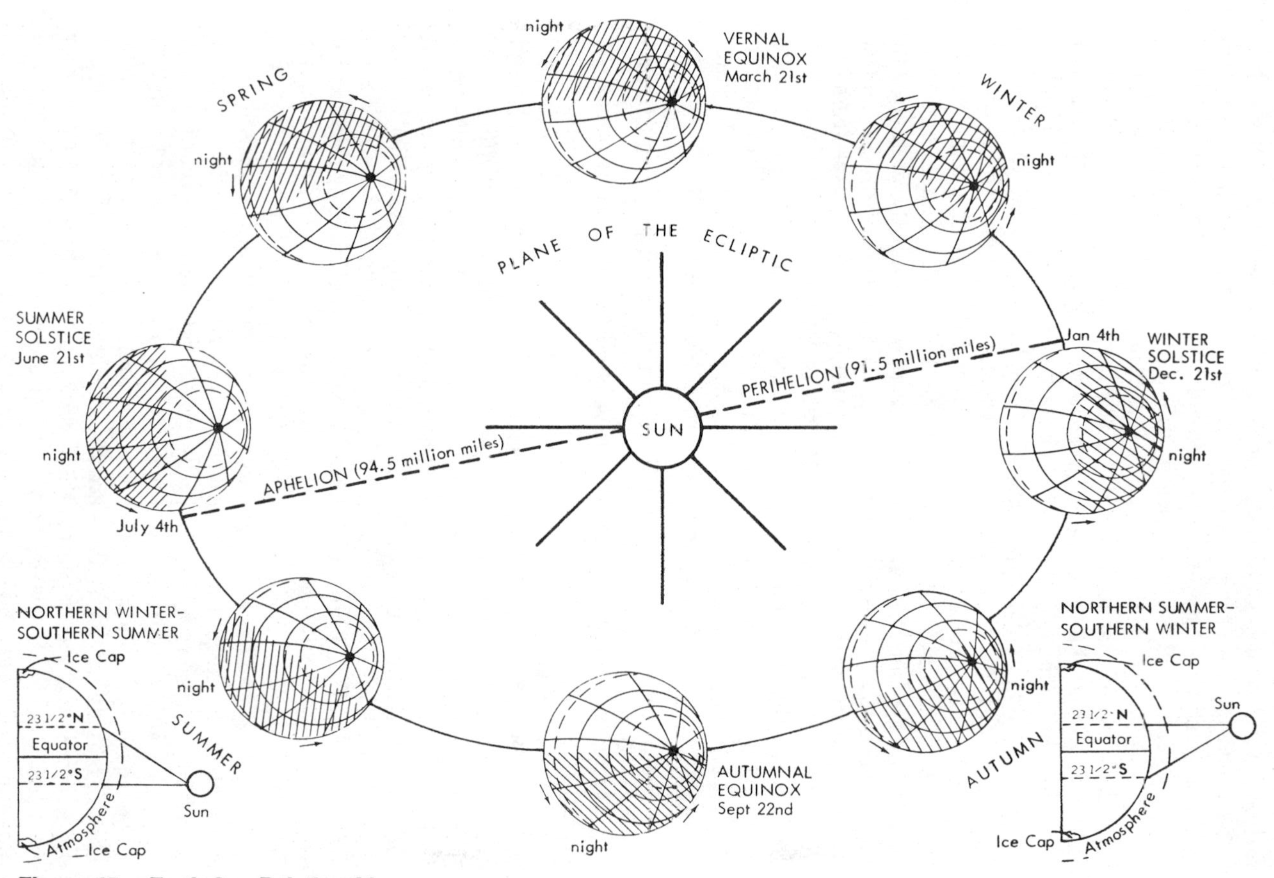

Figure 15. Earth-Sun Relationships.

it is only the surface or surface layers of the land which are involved in seasonal temperature changes. Thus in winter the sea still retains some of the previous summer's warmth and in summer loses the winter's chill very slowly. It doesn't fluctuate in its temperature throughout the year nearly as much as the land. Anyone who lives by the sea or by a large lake has plenty of evidence of this and of the climatic effects of the sea on the land, making it cooler in summer and warmer in winter than areas farther inland. But beyond the littoral zone (the area immediately bordering the sea), these moderating influences may be carried landward by winds; where predominantly sea-to-land winds blow without meeting major topographic obstacles, they tend to create a mild moist climate which is called *maritime* or *marine*, characterized mainly by gentle rainfall fairly evenly spaced through the year and by relatively mild winters and cool summers. This is the climate of northwest Europe, where the great European plain allows the influence of the sea to extend several hundred miles inland. In northwest United States and Canada, however, mountain ranges lie not far back from the coast, and the maritime climate is therefore limited to a rather narrow zone, as it is for the same reasons in southwest South America. Notice on the climatic map that this maritime type of climate is also present in part of South Africa, in southeast Australia, and in New Zealand. This suggests that marine climate is associated with latitude.

Wind Systems and Climate

For reasons which need not be discussed here,[3] there is a more or less orderly global system of winds, with roughly consistent directions in each zone of latitude (see Figure 16). Between about latitude 30° and latitude 60°, both north and south of the equator, the winds tend to blow predominantly from west to east and are called *prevailing westerlies.* Once they reach the land, especially if the land surface is broken up by mountains, their surface direction is much less consistent and may often be reversed, but at altitudes of five thousand feet or more they retain their predominantly easterly flow. If an area lies on the west side of a continent in these latitudes, its climate will therefore be influenced largely by winds of maritime origin. Notice for instance that at latitude 55° N. lie coastal British Columbia, Ulster, Yorkshire, and southern Denmark, all areas of warm winters and cool summers, while at the same latitude in east-coast North America lies northern Labrador and in east Asia, equally frigid Siberia. The maritime climates of southeast Africa and southeast Australia are, however, on east coasts; here the winds are blowing predominantly from east to west. These areas lie far enough toward the equator to be in the zone of the *trade winds,* and east to west is their principal direction. A little farther north on these east coasts, however, where the trade winds are still blowing off the sea, the climate is too much influenced by a strong sun to be considered moderate.

With these two exceptions, east coasts outside the equatorial and tropical zones, where the sun dominates, and outside the arctic, where

[3]These reasons are mainly the latitudinal exchange of heat between high and low pressure areas and the rotation of the earth.

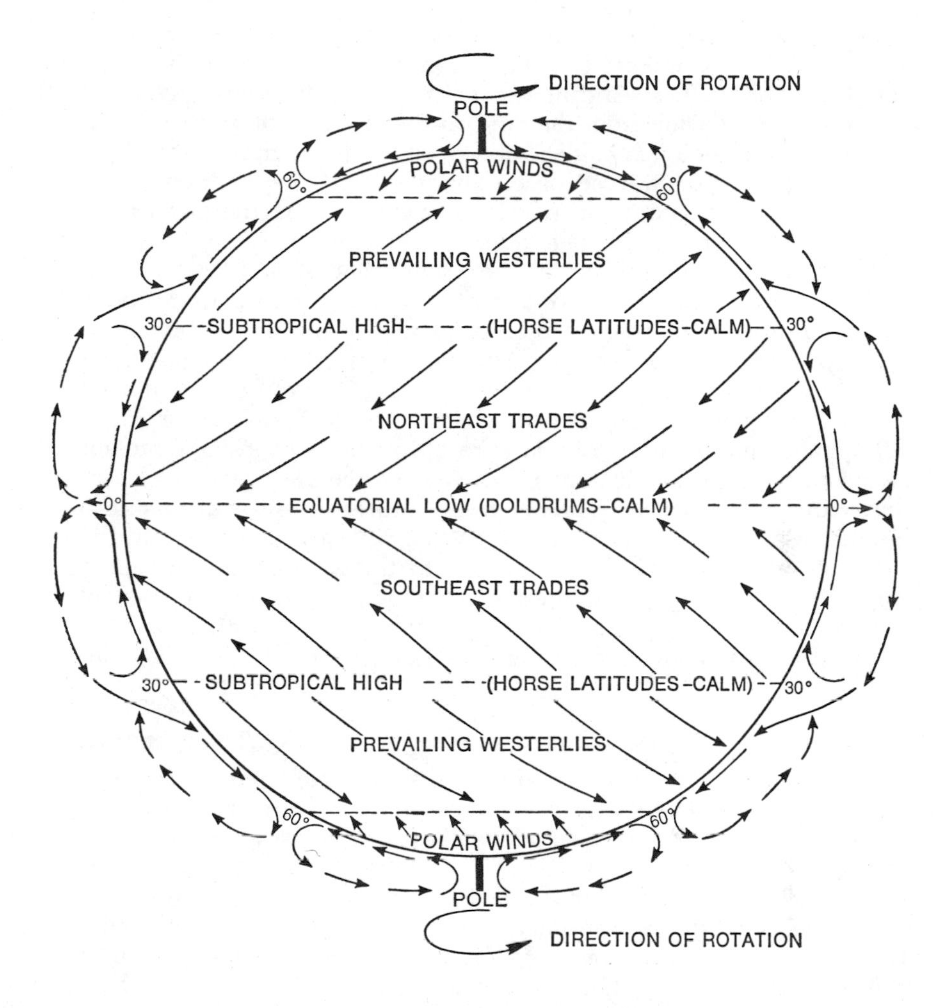

Major wind belts as they would appear on a rotating earth with a uniform surface. Winds are deflected to their right in the northern hemisphere and to their left in the southern hemisphere. Pressure and wind belts move north and south with the sun, according to season.

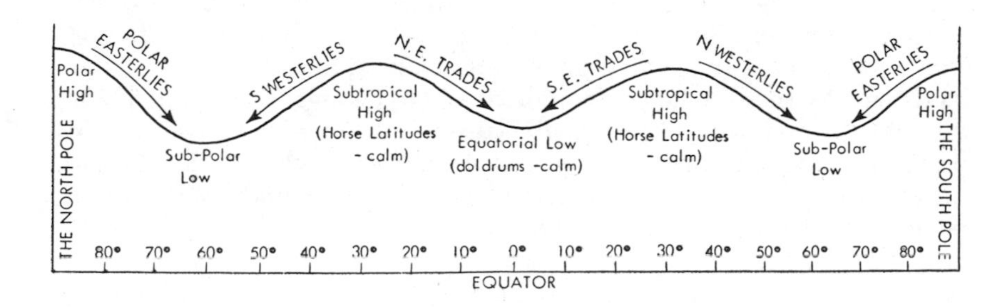

Figure 16. Generalized Wind Belts of the World.

the sun is weak, lie in the path of the prevailing westerlies. Their climate in a sense is imported to them from the great land mass which lies at their backs. The sea cannot affect them nearly as strongly, since the winds are blowing from the land to the sea, and these east coasts consequently have much colder winters and hotter summers than corresponding latitudes on west coasts. The interior of the great continental land masses is an area of temperature extremes, far removed from the sea, where winters are long and cold and summers scorchingly hot. Kansas, Dakota, or Nebraska in the United States, or Soviet central Asia, are typical samples of *continental* climates, with a tremendous temperature range between winter and summer and only a brief part of the year which could be considered moderate between the changing of the seasons. The lowest temperatures, as a matter of fact, were until recently thought to be not as one might expect at the North or South poles, but in northeastern Siberia (the old record was 99° Fahrenheit below zero), because it is much farther from the sea, frozen or open, than either of the poles.[4] Summer temperatures in this area may also be uncomfortably high (unlike the poles), with a July average in the seventies and extremes in the nineties. In the zone of the prevailing westerlies, these continental interiors are the predominant influence on the east coasts, which helps to explain the rigorous winters of New England even though it is far south of London, which has mild winters.

The Monsoon

The fact that the continental interiors have such marked seasonal temperature differences also contributes to the *monsoon*. In summer heated air rises over the center of the great land masses, and cooler air is drawn in from the surrounding sea to take its place. If this marine air is cooled after it strikes the land, especially if it is forced to rise by mountains or hills, or if it meets a cold air mass, heavy rainfall results. In winter the main direction of flow is reversed. The cold, dense air of the interior sinks, and winds blow out toward the sea, bringing little or no moisture with them since they originate in a very dry place and are cold to start with. The rainfall associated with the inblowing summer monsoon rarely penetrates into the interior; the winds are milked dry by mountains or lose their moisture or their force in other ways before they reach the great interior deserts. Monsoonal winds are most clearly developed in the largest land mass, Eurasia, as one would expect, where the pull of rising heated air in summer and the push of falling cold air in winter are particularly strong. Only in the Asian part of this land mass is there a large enough area far enough from the sea to establish this dominant pattern. In Europe the sea is too close, but in parts of east Africa, where the continent bulges widest, some monsoonal effect develops, although it is usually dominated by the more powerful and adjacent Asian monsoon.

---

[4]During the International Geophysical Year in 1957, a low temperature of 102° Fahrenheit below zero was recorded at the South Pole, and in 1958 the Russian base in Antarctica recorded a new low of 125° Fahrenheit below zero.

In the other continents monsoonal patterns are not readily apparent, principally because they are not big enough land masses, but the differential heating and cooling between land and sea do allow marine influences to penetrate farther inland in summer than in winter, especially on east coasts. This is part of the reason for the hot, sticky summer days in places like Washington, D.C., or New York; some of their uncomfortable summer weather is imported from the tropical Caribbean. Even in Asia, however, the actual forces involved in the weather systems which are broadly labelled monsoonal are very much more complex than this simple account suggests and are by no means fully understood by meteorologists. It has been said that "every schoolboy understands the monsoon, but the Indian Meteorological Service is still puzzled by it."

The Seasons

Seasonal changes result from the 23.5° inclination of the earth on its axis, as illustrated in Figure 15. The sun's position in the sky in relation to the earth, and hence the angle at which its rays strike the earth, migrates annually north and south of the equator. The sun appears directly overhead at noon, and its rays therefore reach the surface of the earth directly, only between the Tropics of Cancer in the northern hemisphere and Capricorn in the southern hemisphere. June 21 marks the farthest point of the sun's apparent northerly migration, when its noon rays are directly overhead at the Tropic of Cancer, 23.5° north of the equator. On December 21 its southerly migration reaches the Tropic of Capricorn at 23.5° south. The zone between these two lines, or the tropics, thus receives maximum insolation which varies little throughout the year and is continuously warm with little or no seasonal changes in temperature. Beyond the tropics insolation and the relative lengths of day and night not only vary from season to season but heating effectiveness is greatly affected by the angle at which the sun's rays penetrate the atmosphere. The atmosphere acts as a powerful insulating blanket. The difference between the result produced by vertical insolation and by insolation at an angle is great enough to account for a large part of seasonal changes. This is readily demonstrated by observing the difference in heating effectiveness between the sun at noon and at dawn or dusk.

CLIMATE AND ECONOMIC OPPORTUNITY

This highly generalized summary of the mechanics of climate should not discourage examination of climatic processes and regional differences in more detail. A good starting point is provided in any map of world climates. But this discussion is concerned more importantly with the effects of climate on human society. Having discarded the thesis of climatic determinism, and having agreed also that at the extremes, such as the desert or the arctic, climate's effects are too obvious to require discussion, what scope is left for the climatic factor? As already suggested, it may be clearest and most accurate to think of climate in terms of the kind and degree of economic opportunity which it offers. This concept most nearly fits the correlation between climatic areas and areas of human settlement or of economic activity and does not inveigle us into considering the effect of climate on man's mind.

In Monsoon Asia

The climate of monsoon Asia, for example, offers an opportunity for intensive agriculture based on wet rice. In this area of long growing season, high summer temperatures, and generally adequate but not excessive rainfall, rice can yield more food (or calories) per acre than is produced by any other cereal crop or by any other area of comparable size. The connection between this climatic advantage and the very large and dense populations of monsoon Asia is clear, in terms of economic opportunity. Irrigation makes almost any agriculture more productive. Monsoon Asia's rainfall is seasonally concentrated in summer and is highly variable from year to year in amount and in seasonal incidence, so that irrigation is especially necessary. Rice is the dominant crop, which means high water requirements. Irrigation became a fundamental basis of most agriculture in monsoon Asia, and the complex political, administrative, and economic organization necessary to maintain irrigation systems has contributed to the general nature of most Asian civilizations, especially to the dominant role of the state, or at least of corporate groups, and the relatively weak position of the individual entrepreneur. The productive nature and special form of their agricultural systems lie at the root of all the great Asian civilizations and have done much to influence their character and to distinguish them from other civilizations elsewhere. Climate has not made the Asian's mind work differently from the European's, but it has provided him with a different kind of economic opportunity. This in itself tended to encourage the growth of a distinctly Asian culture. Climate was not of course the only factor, but in this case its role, in a permissive sense, is apparent.

In the Steppe

The semi-arid areas of the world are in a different climatic situation. *Steppe* is used loosely as both a climatic and a vegetation term. Climatically it applies to the semi-arid areas, transitional between desert and humid, too dry for any significant tree cover but moist enough to support grasses or scrub. Steppe covers the Great Plains of the United States, most of the Prairie Provinces of Canada, part of the Argentine *pampa*, a broad rim around the great Australian and African deserts, and most of middle and southern central Asia. Here climate offers a relatively limited economic opportunity, and of a particular kind. Except in the relatively small areas where irrigation is feasible, or where there are particular local opportunities such as minerals, the steppe does not support a large or dense population. Most of the people who live there earn their living *extensively*, by applying relatively small amounts of labor or capital to proportionately large amounts of land. The climate, except in special situations, does not make a heavy investment of labor or capital profitable because the return would not be great enough to compensate for the expense, as it is for instance in monsoon Asia or New Jersey, where land is worked *intensively*. Steppe areas therefore tend to be characterized by sparse populations engaged in growing drought-tolerant crops like wheat on very large farms in the best or least dry sections, or in the raising of animals—cattle, and in the drier or hillier parts, sheep or goats.

If mineral resources are fortuitously present, mining may supplement the economy and attendant cities may grow up as trade, transport and service centers. Denver is a good example. Denver also profits from location where plains give way to mountains and where transport routes bunch together to get across the easiest route. Bulk is commonly broken there, and Denver makes a good assembly and distribution center, as well as a good operations base for railroads and airlines, all of which provide basic employment for Denver's population and in turn support local service industries and service employment. Denver would be a much smaller place, however, if it were not that it lies between two much more densely settled and productive but complementary areas, eastern United States and the west coast. Exchange between them takes place across the intervening steppe, through Denver. The large amount of intervening space, plus the barrier of the mountains, makes a trade service center and break-in-bulk point useful along the east-west route. The same set of factors is involved in the pattern of trade service centers across the steppe of Eurasia and on its margins. But manufacturing in the steppe is unlikely, aside from the primary processing of local raw materials such as ores or the supply of the small local markets. The major markets will be elsewhere, where the people live, in better watered places.[5] The steppe is also likely to experience considerable climatic fluctuation, especially in rainfall, since high variability is unfortunately associated with low annual averages. In a commercialized economy like the United States, fluctuating rainfall may be accompanied by waves of prosperity and depression in the steppe, with attendant political unrest or support for radical policies of relief or reform. In a precommercial or nomadic economy such as central Asia's has been until recently, climatic fluctuations may contribute to conflict among nomadic groups, or between the nomads and the sedentary farming cultures around them.

Climate and Rail Patterns

In the United States the pattern of transport routes reflects this climatically induced economic nature of the steppe. A map of railroads makes this point unmistakable. At about the hundredth meridian of longitude, the rail pattern in the United States contracts more or less abruptly and sends west into and across the plains only a few lines, in sharp contrast to the dense network to the east. At first sight this might appear to be due to the mountains, especially if one looks at a generalized physical map of the United States. Actually, the area concerned is the Great Plains, and while there is a rise in elevation westward from the Mississippi, it is so gradual that the landscape looks virtually flat or

[5]The Soviet Union has, for strategic and political reasons, located considerable manufacturing in Russian steppe areas.

gently rolling, and there is no real barrier of hills or mountains until the sharp front of the Rockies is reached. This is one of the disadvantages of the altitude tint system of showing elevations, which is discussed in Chapter 12. On a rainfall map, however, a major break is shown which coincides quite closely with the hundredth meridian. This is the line (actually a zone) between steppe (which merges with a more humid but still treeless area called the prairie) and the well-watered east, between two areas of differential economic opportunity. The line of twenty inches average annual rainfall, or the twenty-inch *isohyet,* is one of the most important regional lines in the United States. Population reflects it, with relatively dense settlement to the east and sparse settlement to the west.

These are matters of great importance for a railroad. Transport routes in general, as pointed out in the analysis of Switzerland, react far more sensitively to the volume and direction of trade than they do to physical barriers. A railroad represents a heavy capital investment, and such investment does not pay unless there is a large and constant volume of business. There are no important physical obstacles for a railroad in the Great Plains, but there is a clear economic problem—a small volume of commodity flow, since there are relatively few people and few large cities. Climate is largely responsible for this. The railroads in effect contract their routes and get over the empty space as economically as possible by what are called "bridge lines." The two ends of the bridge are the humid east and the humid or irrigated Pacific coast, where there are enough people and enough exchange between them to carry the railroads across the intervening relatively empty space.

A similar contrast can be seen in Australia, for the same reasons. Here the east-west bridge lines have contracted to only one. Interior Australia is a true desert, as the place names emphasize (including the area known as the "Nullarbor [no trees] Region"), and its emptiness almost excludes railroads entirely. The two ends of the Australian bridge are also much smaller and less productive than eastern or Pacific coast United States and hence less able to maintain bridge lines between them. In addition, the two humid coasts are easily and cheaply linked by sea transport, cheaper than any form of land transport, and the railroad is there at all largely for political and strategic reasons. There is no central north-south line; the economic inertia of the desert, here at its widest and driest, is too great. The two ends of this incomplete bridge are too small, in terms of population or complementary exchange (the southern end is a desert, and the northern end lies in the hot tropics), to make the connection pay.

On a map of North American railroads, Canada seems to deny what has been said about the hundredth meridian. To the west of it in Canada the railroads actually spread out, whereas to the east they contract into only two lines, bridging what is obviously an economically empty space before they fan out again into the humid areas of eastern Canada. Climate is responsible for only part of this, but since it has come to our attention at this point, it makes a useful demonstration not only of the climatic factor but also of the fact that the environment tends to operate as a whole as an influence on mankind. Look first at the empty space. That part of Canada is an area of nearly bare crystalline rock known as the Canadian Shield; most of the original soil was scrapped off by the last glaciation, and what little remains is too thin and poor to support anything but a spruce-fir forest except for a few limited areas. It is an area largely avoided by human settlement, as the map of population shows, but soils are more important than climate in accounting for this. Climate in the Canadian Shield is somewhat less good for agriculture than in the Prairie Provinces farther west, despite the fact that the average annual rainfall is higher in the Canadian Shield; the rainfall is concentrated in late summer, which means that it is less useful, and the summers as a whole are shorter and cooler than in the Prairie Provinces. This is minor, however, compared with the virtual absence of usable soils. Actually there are large and valuable deposits of mineral ores in the Canadian Shield: nickel, iron, copper, zinc, uranium, and others. They do not provide a basis for large-scale settlement because, in the absence of other economic opportunity in this generally marginal area, the ores are shipped out to where the people live and where the major markets and manufacturing centers are. The railroads thus have to cross this empty space of the Canadian Shield as economically as possible, picking up some ores on the way, but not really running on a profitable basis until they reach southeastern Canada, where soils, climate, location, and other factors combine to provide economic opportunity for denser human settlement.

West of the shield lie the Prairie Provinces—Alberta, Saskatchewan, and Manitoba. Here the map shows markedly greater population density than in the shield (or than at corresponding longitudes in the United States to the south), and a spreading out of the rail routes to service it. Climate has played here both a direct and an indirect role. Directly it is relevant because, while average annual rainfall is much the same for the Prairie Provinces as for the Great Plains area of the United States (which has a much sparser population), it is more effective because evaporation rates are lower in this cooler climate. It is thus possible to grow wheat, for example, on a larger basis in Alberta than in Wyoming or Colorado. A map showing generalized types of farming makes this clear and shows a skewed pattern of wheat farming extending much farther into the dry west on the north than in the south, leaving most of the Great Plains in the United States dependent on grazing, and hence supporting relatively fewer people.

Indirectly, climate has helped to account for this difference in terms of alternate opportunities. Canada has few favorable alternatives to the Prairie Provinces in terms of general economic opportunity, whereas the United States has the Middle West as an obviously superior domestic rival agriculturally to the Great Plains, and a wealth of other areas capable of supporting dense settlement on a high economic level. Most of Canada is too cold (or in the case of the shield lacks also the basic factor of workable soils) to permit large-scale settlement, and the climate of the Prairie Provinces makes it one of Canada's favored areas where it might be a poor one in the United States and would be used or settled much less fully.[6]

Climate and European Settlement in Africa

There are some four million Europeans settled permanently in Africa south of the Sahara. With few exceptions, their settlements are confined to two types of areas: highlands and coastal or river ports. It is frequently asserted that the Europeans live in the highlands because the climate of most of Africa prohibits them from living elsewhere, that it saps their energies, destroys their wills, and makes it physically impossible for them to keep going at their accustomed European pace. There is no doubt that most lowland African climates are highly uncomfortable for someone who comes from a climatic situation like Europe's. But Europeans do live there, in the coastal and river ports, and they live in climates which are equally uncomfortable to them in places like Manila, Singapore, or Calcutta. These places offer Europeans economic opportunity, principally in trade. The Europeans complain about the climate, use it as an excuse for special salary, housing, vacations, and diet, but they live and earn their living in these hot lowlands with obvious success. They do not live to any extent in the lowland areas outside the cities. Here there are few essential trade functions for them to perform, and the climate offers them little economic opportunity.

The European goes abroad in order to make money. He could make little in competition with local agriculturists in a tropical climate, except as a manager of a plantation or a trader in its products. Up on the highlands, however, the climate offers him a basis for commercialized agriculture of the European type, mainly cereals and commercial stock raising for export, and it is on this that the majority of European settlements in the African highlands have been built. Later discoveries of gold and diamonds or of other mineral resources have cemented the European occupance and made it more profitable, but the European is there because he can make an attractive living, not because he finds the climate personally or physically pleasant.

Climate and the Development of Europe

Northwest Europe's climate has a mild, maritime character and generally lacks the extremes which in other parts of the world impose strict limits on what man can do. But there are certain clear positive advantages. Agriculturally, Europe's climate is among the most advantageous in the world, its main drawback being the coolness of its summers and the shortage of strong sunshine; this largely excludes sun- and heat-loving crops like maize or cotton, and of course the tropical crops like rubber or sugar. But no single climatic area can hope to be all

---

[6]There is a non-climatic factor at work in this part of Canada which will continue to increase its population and prosperity still further; large deposits of petroleum and natural gas (in addition to existing coal) have been discovered and are being worked there. All three are the most important deposits in Canada as a whole, although they do not rival the large United States deposits.

things to all men. Northwest Europe's adequate and gentle rainfall evenly distributed throughout the year, its long growing season and mild winters, rank it high as an agricultural area and helped to make it possible for Europe to support the agricultural and industrial revolutions, and to support a relatively large and highly developed population.

Northwest Europe still contains several specialized and highly profitable agricultural areas, but its principal economic activities are trade and manufacturing. It may seem difficult to relate these to climate, and yet there is a relationship. The volume of trade is in part dependent on the size of the economic base: the more people, or the greater the production, the greater the trade, assuming other factors to be favorable, such as cheap exchange, the complementary nature of the regions involved, and the commercialized nature of the economy. There are lots of people in China and lots of production but, until recently, proportionately little trade, because most of these factors were not favorable. In Europe they are preeminently so; the permissive nature of the climate allowed or encouraged the development of a large base of population and production. Climate also contributes to complementarity between the regions of Europe and between Europe and the rest of the world. In part because its climate excludes the profitable growing of maize, northwest Europe has an extensive trade with Argentina and North America, importing cereals and animal products (raised on maize) and exporting manufactured goods. In part because the Mediterranean Basin has a drier and warmer climate than north Europe, the two areas exchange their different products, as was noted in the analysis of trade routes through Switzerland. Manufacturing tends to concentrate in areas where raw materials can be cheaply assembled and where finished products can be cheaply distributed to the greatest number of customers. In both respects, the most advantageous location is likely to be in or near the major centers of population, where plenty of labor, including skilled labor, is also available. Mineral resources in the steppe or in other generally unfavorable climatic areas do not create major manufacturing where they are. Instead, manufacturing occurs in the market centers. Climate plays a part in influencing the economic opportunity which accounts for the great market centers or the great clusters of population and hence helps to support Europe's specialties of trade and manufacturing.

Climate, as repeatedly pointed out, plays a predominantly passive or permissive role. It allows certain developments rather than causing them. Accessibility, cheap exchange, local mineral deposits, economic and political organization, the availability of capital surpluses, and the growth of technology are more important than climate as advantages for trade and manufacturing. Northwest Europe remained a backward area until these other advantageous factors were developed or became effective, although the climate did not change. The relevance of the favorable climate is that without it, the other advantageous factors would be less effective or even inoperable. There is at least one respect, however, in which northwest Europe's climate has been a more direct advantage for trade and manufacturing. Cheap exchange, or cheap assembly and distribution, is a crucial factor for both of these functions. The mild, moist, maritime climate of Europe ensures that the great European plain will be traversed by many rivers, and that their navigability will be enhanced. In maritime Europe, west of Poland and south of Norway, rivers do not often freeze solid in winter, and serious floods or prolonged low-water periods are relatively infrequent, since the rivers receive rainfall quite evenly distributed throughout the year. The Rhine, for instance, is a relatively short river, but it is navigable nearly up to the mountain front of the Alps, including a short navigable stretch within Switzerland itself. This has been an important asset for a Europe so heavily dependent on cheap exchange. Water routes, by sea or by river, join together nearly all of its parts and make possible cheap movement of bulky low-value goods, like coal or foodstuffs, which are essential in large volume to maintain a modern specialized industrial economy. Internal waterways helped to make the industrial revolution possible in Europe before the building of railways, and after industrialization water routes could move bulk goods more cheaply than railways. Rivers remain a vital economic asset for Europe, and their usability is directly related to Europe's climate.

This chapter has attempted to discuss climate as an influence on human society primarily in terms of economic opportunity. It has been necessary to generalize very broadly in order to make the essential point and to focus attention on climate to the disregard of other factors. Climate, like any other influence on man, does not operate alone, but as part of a complex in which the rest of the environment, relations in space, and other separate or non-geographic factors are also involved. Climate is one of many factors lying, one may say, at the foundation of man's development and giving him the material, good or bad, with which he is obliged to work.

Questions for Further Study and Discussion (Chapters 7 and 8)

1. What is the length of the growing season in the area where you live? How does this affect local agriculture? Has it any other effects on human settlement?

2. Is the climate of your area marine or continental, or can it not be clearly labelled as one or the other? Why?

3. How does climate affect interregional exchange?

4. How much of the precipitation in the area where you live falls as snow, sleet, or hail? How much do all of these total as rainfall equivalents?

5. Has man conquered nature?

6. Collect if you can statistics or information on climatic fluctuations over the world as a whole, or in the area where you live, during the past one or two hundred years. What do you conclude? (Consult the Landsberg and Shapley entries in the reading list for this chapter.)

7. Are there places other than China and India which lie wholly or in part in the tropics and which have also developed civilizations which were technically or culturally more advanced than their contemporary cultures in cooler climates?

8. Why should a productive agricultural base, in Europe or elsewhere, be a factor in industrialization?

9. How is the size of Denver related to the distance between St. Louis and San Francisco? Denver is said to have the largest hinterland of any city in the United States. Why? Why then is Denver not a larger city?

10. Where can one find dissimilar civilizations in similar climates? Similar civilizations in dissimilar climates? Why in each case?

11. What specific detailed information would you want about the climate of an area in planning the development or extension of an agricultural system? Why is each piece of information necessary?

12. What is the predominant type of rainfall in hot deserts? Why? What significance is this likely to have for human settlement?

13. Denver is the largest steppe city in North America, but Kiev and Kharkov in the Russian steppe are much larger. Why?

14. The rubber tree (*Hevea brasiliensis*) will grow well over a very large area of the hot, wet tropics where climate may be assumed to be more or less uniform. Actually it is grown commercially in only a few small areas. Why?

15. The island of Java is very much more densely populated than the other islands of Indonesia. Is this difference due to climate? If not, what is it due to?

16. What is meant by *micro climate?* How does it differ from *macro climate?* What is the significance of these differences? (Consult the Landsberg article in the reading list for this chapter and its references to other studies.)

17. What is a diurnal monsoon?

Selected Samples for Further Reading (Chapters 7 and 8)

Barry, R. G., and Chorley, R. J. *Atmosphere, Weather, and Climate.* 2nd ed. London, 1971. A good general text, technical and theoretical as well as empirical.

Bates, M. *Where Winter Never Comes.* New York, 1952. A general and semi-popular but careful account of the tropics and some of the human adjustments to climatic conditions there.

Blumenstock, D. *The Ocean of Air.* New Brunswick, N. J., 1959. An excellent general treatment of meteorology, with special attention to problems of weather prediction and control and the effects of climate on man.

Brooks, C. E. P. *Climate through the Ages.* London, 1949. A useful, clearly written survey of climatic change through geologic and historical time.

Chang, J. H. *Climate and Agriculture: An Ecological Survey.* Chicago, 1968. A useful overview of the effects of climate on plant growth and the consequent problems of the farmer.

———. "Progress in Agriculture Climatology," *The Professional Geographer* XX (1968), 317–20.

———. "Agriculture Potential of the Humid Tropics," *The Geographical Review,* LVIII (1968), 333–61. Two important studies of techniques for measuring and evaluating climate, and the need for more careful adjustment of farming practices to the demanding and only marginally rewarding agricultural environment of the humid tropics.

Gourou, P. *The Tropical World.* Trans. S. H. Beaver, 4th ed. London, 1966. The nature of the tropical environment and the variety of man's adjustments to it.

Hare, F. K. *The Restless Atmosphere.* London, 1963. Also available in paperback, this is a brief, clear survey of meteorology.

———. "The Westerlies," *Geographical Review,* L (1960), 345–67. A technical but clear discussion of some of the recent findings and further clarifications about the world pattern of air movements in the zone of the westerlies, including the importance of "whirls" and their relation to monsoonal air movements.

Huntington, E. *Mainsprings of Civilization.* New Haven, 1945. The last work of a noted climatic determinist, in which he attempts to summarize his findings but fails to make a convincing case.

Kendrew, W. G. *The Climates of the Continents.* Rev. ed. London, 1961. A standard work summarizing data on regional climates.

Koeppe, C. E., and De Long, G. C. *Weather and Climate.* New York, 1959. A recent standard text.

Landsberg, H. E. "Trends in Climatology," *Science,* CXXVIII (1958), 749–58. An excellent concise summary of knowledge in this broad field, with an indication of future possibilities.

Lee, D. H. K. *Climate and Economic Development in the Tropics.* New York, 1957. A careful analysis of this complex problem, including some consideration of climatic effects on individual capabilities.

Manley, G. "The Revival of Climatic Determinism," *Geographical Review*, XLVIII (1958), 98–105. A useful review of the changes in the popularity of climatic determinism, notes on some sample historical studies, and some suggested conclusions.

Murphey, R. "The Decline of North Africa: Climatic or Human?" *Annals of the Association of American Geographers,* XLI (1951), 116–32. Climatic versus non-climatic factors as explanations of changes in the settlement and use of a semi-arid area.

Patton, C. "Professional Contributions to Physical Geography." Chapter III in P.E. James, ed., *New Viewpoints in Geography.* Washington, 1959. Includes an excellent, brief, clear survey of some recent climatic theories.

Pedelaborde, P. *The Monsoon.* Trans. M. J. Clegg. London, 1963. Clear introductory survey of the current state of knowledge of a complex meteorological pattern.

Rumney, G. R. *Climatology and the World's Climates.* New York, 1968. A good standard textbook.

Sauer, C. "The End of the Ice Age and Its Witnesses," *Geographical Review,* XLVII (1957), 29–43. A sample of the literature on the fascinating subject of early man and his development in relation to the great climatic changes which he survived.

Sewell, W. R. D., ed. *Human Dimensions of Weather Modification.* Chicago, 1966. Discussions by specialists from different fields about the technical possibilities of modifying some aspects of local weather, and some of the possible consequences.

———, Kates, R. W., and Phillips, L. E. "Human Response to Weather and Climate," *The Geographical Review,* LVIII (1968), 262–80. A detailed review of the spread of geographical literature on this absorbing topic.

Shapley, H., ed. *Climatic Change.* Cambridge, Mass., 1954. A collection of articles by leading authorities on the numerous aspects, mechanics, and possible consequences of climatic change.

Trewartha, G. T. *An Introduction to Climate.* 4th ed. New York, 1968. A good semi-technical general survey of the physical aspects of climate.

United States Department of Agriculture. *Climate and Man.* 1941 Yearbook of Agriculture. Washington, 1941. This large volume includes an excellent collection of general articles by leading authorities on climate and human settlement, climate and agriculture, and problems of forecasting, with an extensive section of climatic statistics for the United States and the world. Some of the material, however, is becoming out of date.

9
Plants and Man

VEGETATION IS LARGELY a corollary of climate, but it can much more easily be modified. It is a useful classifier, since it reflects the climate and may also reflect soils, ground water, and relief, as well as the activities of man. One of the best kinds of evidence of past climates and often of past land use is the plant pollen deposited or fossilized in the soil. Vegetation also reflects the nuances and variations of climate from one small area to another and from season to season or year to year, so that it may often be a more reliable guide to climate than is provided by short-term records. Comparison of a world map of vegetation with a climatic map will demonstrate how closely the two co-vary in their spatial distribution. The vegetation map shows, as it indicates, "natural" vegetation, however, and in the parts of the world where man has settled in large numbers most of the original vegetation cover has been removed. The nature of the original vegetation cover has had to be reconstructed on the basis of climatic and soil conditions, plant pollens, and by means of the few surviving specimens. In some areas where man has not settled so densely, the original vegetation largely remains and is in part responsible for restricting human settlement. The most obvious and important example is the equatorial rainforest of Africa, South America, and southeast Asia, where the barrier effects of vegetation help substantially to limit economic opportunity. Vegetation, however, does not act alone. The climate of the equatorial rainforest is harsh, the soils are generally poor, and the areas as a whole cannot support enough people at a high enough economic level to make the removal of the rainforest physically or economically possible except on a relatively small local scale. The same limits apply to most of the wide band of coniferous forest in the northern hemisphere between about latitude 50° and latitude 70°, known as the *taiga*.

BARRIER FORESTS

Apart from the useful or cultivated plants, vegetation tends to act as a physical barrier in restricting human activity, although where trees are absent and grass is dominant, vegetation may offer an opportunity for movement. As already pointed out, where vegetation has been a barrier in the densely settled areas, it has largely been removed. But before this removal was accomplished, or before other conditions made the removal physically and economically possible, forests especially were an important limit on human activities. The role of the forest in European history was that of an obstacle to trade and circulation, to agriculture, and to the achievement of national unity. The removal of most of the great European forest stands at the end of the Middle Ages coincided with a great increase in trade, with the growth of towns, and with the rise of modern European civilization. The forests also acted as refuge areas for persecuted or dissident groups and as bases for outlaws or robber bands. Sometimes the forests served to shelter more organized attempts to levy tolls on traffic passing through or around them and were the seats of the so-called robber barons. Most historians believe that the story of Robin Hood is a cherished fiction, but at least there were people like him in Sherwood and in many other forests in England and Europe. Sherwood forest is almost gone now; the area where it once stood is in the modern English industrial belt. Nottingham Castle still stands, but it is surrounded by a manufacturing city, and tourists are taken to an obviously cut-over and tiny patch of forest to see the "Major Oak" under which Robin is supposed to have assembled his band. When trade and its urban bases grew large enough, and when the national states grew strong enough to put down local anarchy, most of the lowland forests of Europe were destroyed, although in the remote or hillier areas some extensive stands remain and are used as sources of timber or for recreation. An equally if not more important factor in the removal of the forests was the expansion of agriculture to feed the growing towns and to support an increasing rural population.

In other, heavily forested parts of the world, the forest still shelters dissident groups. In Malaya, for example, it has been said that "the jungle is not neutral; it is immensely biased against the forces of law and order. The fugitive, the outlaw, the guerrilla fighter, has every advantage."[1] During the war in the South Pacific theater from 1941 to 1945 the forest was a military factor of supreme importance.

In Europe, surviving place names record the once nearly universal forest cover and the long struggle against it—for example, English names like Brentwood (burned wood) or names with endings in -ley (clearing), or German endings like -wald (wood), -holz (wood), or -hau (cut down). In a few places where the local political and economic order was upset or the population decimated, the forests recovered and even grew over once-cultivated ground, as in large parts of Germany after the Thirty Years' War. There has been a continuous rivalry between man and the forest, in which the forest has often been associated psychologically with the unknown, the dark, magic, the home of feared spirits, and the symbol of mystery.

---

[1]Vernon Bartlett in the *Manchester Guardian Weekly*, Nov. 22, 1956, p. 15.

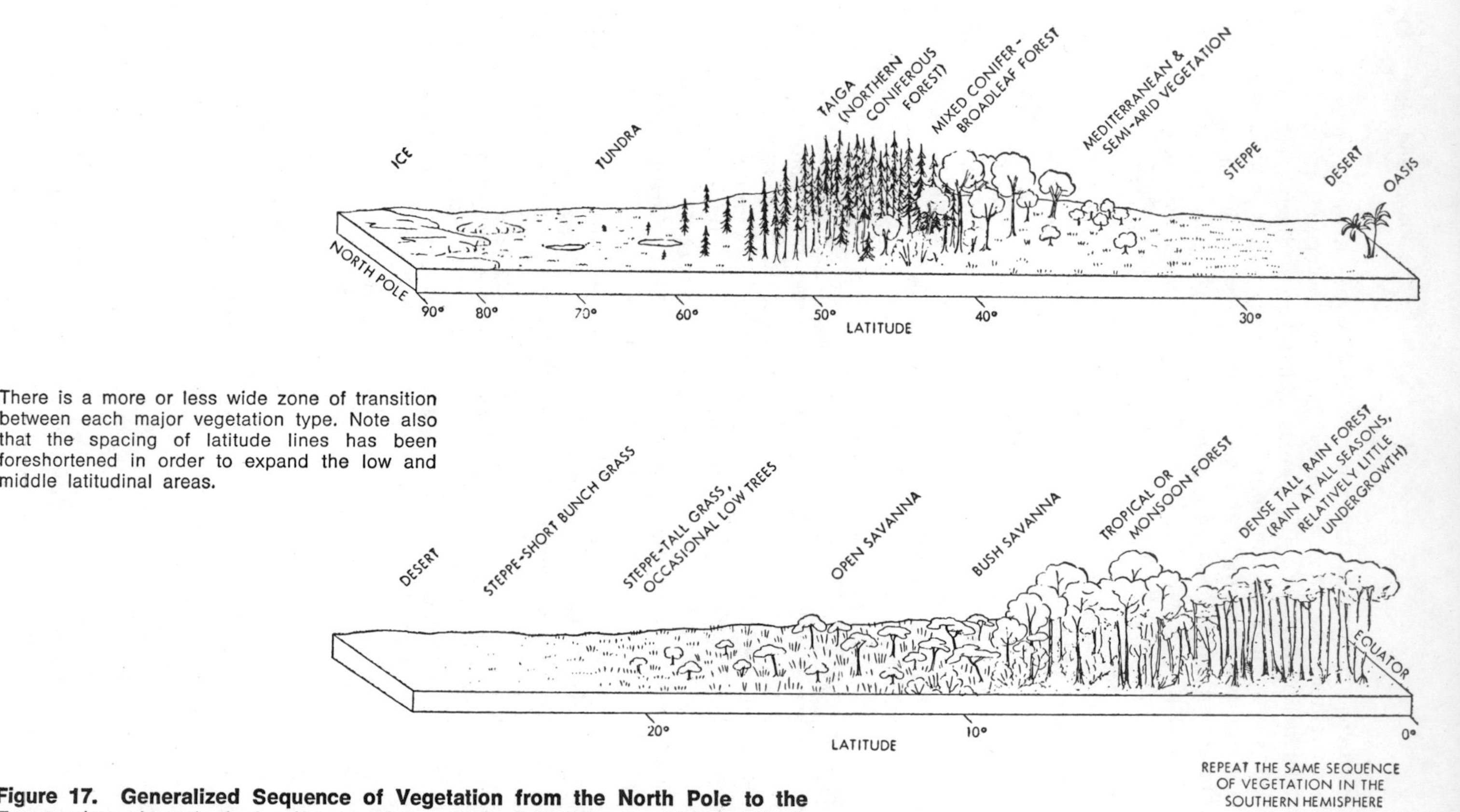

There is a more or less wide zone of transition between each major vegetation type. Note also that the spacing of latitude lines has been foreshortened in order to expand the low and middle latitudinal areas.

Figure 17. Generalized Sequence of Vegetation from the North Pole to the Equator (on a hypothetical uniform surface, eliminating differences in elevation).

Forests in History

One famous European forest stand, long since razed, has nevertheless left its mark behind. This is the Carbonnière forest, which until the fourteenth century lay across what is now northern Belgium. Its importance to the local or nearby people is suggested by the fact that it shares its name with the French words for coal and charcoal (which was used long before coal), and with the English word carbon. But the Carbonnière forest had more far-reaching effects. Its site coincided with a cultural, ethnic, religious, and linguistic boundary in Europe, between the Latin south and the Teutonic north. Northern Belgium is mainly a flat plain, and it is puzzling to find running across the plain such a major cultural line. The Belgian national state is a recent creation, but long before it came into existence and until the present, people living there south of the area of the Carbonnière forest spoke French or one of its dialects and belonged in other respects with the Latin half of Europe, whereas to the north they spoke a Germanic language and belonged in the Teutonic or Germanic camp. These areal differences, reflecting in part the original barrier of the Carbonnière forest, provided one of the main bases for the creation of the Belgian state in the nineteenth century, five hundred years after the forest had disappeared.

The earliest civilizations arose in largely treeless areas–Egypt, Mesopotamia, north China, and northwest India. Before recorded history man seems generally to have avoided dense forest, at least for his settlements. Neolithic settlements in Britain, for example, were concentrated on the open chalk downs, probably because the original tree cover there was sparser than on the heavier clay soils and was more easily removed. Once man was able to use and manipulate fire, perhaps as long as half a million years ago, he had a powerful weapon against the forest or against any vegetation. He also learned to use stone tools to kill trees by girdling or felling them. Human agency has thus altered vegetation on a large scale for a very long time. Burning was used to clear forests of underbrush in order to make hunting, grazing, or movement easier, or to improve pastures by favoring the types of vegetation which flourish after fires. Forests have seldom been attractive to advanced human settlement. No heavily forested area, so far as we know, has ever been the base for a technically or economically advanced civilization, with the possible exception of the Maya culture of Central America, whose origins are obscure. It seems likely that there, as in other forested areas, sophisticated development took place only after the trees were removed.

Rivers have often been the first avenues of attack on the forests and the first corridors of denser settlement, as the only zones of easy movement within an otherwise barrier area. In Europe and Asia deforestation began along the rivers, as it did and is still doing much later in the other continents. Forests were cut into for practical uses as well as to clear the land for cultivation or trade. Wood has always been in demand as building material and fuel, and for charcoal as a more efficient fuel. Charcoal burners were often the pioneer forest destroyers, ranging widely and reducing the wood to charcoal in small kilns within the forest so as to use local fuel and to lessen the transport problem by converting a bulky, low-value product into a more compact, higher-value one. As the ravages of the charcoal burners progressively exhausted the more accessible timber stands, industries dependent on charcoal suffered, especially the smelting of ores. This problem was an important factor in the rapid use of coal on a larger scale, in the form of coke, in eleventh-century China and in seventeenth- and eighteenth-century Europe. In modern northwest Europe, deforestation took place earliest in England. In addition to the fuel problem, there was a shortage of ship timbers. Oaks large enough for the keels and stern posts of naval ships of the line were scarce by the eighteenth century. The development of metal ships came in time to solve the problem, but for a considerable period the size and number of capital ships had been to a degree limited by the decreasing availability of large oak timbers. Another important use of forests in Europe was for the grazing of pigs on beech mast (beech nuts); this represented a significant addition to food production, especially where open beech forests were accessible to densely populated areas.

Although the original forest cover of Europe was modified and reduced by Neolithic and Paleolithic man, it was still extensive enough to help in limiting the area of the Roman conquest. The Roman military and cultural impact on Germany and the more heavily forested parts of central Europe was slight. In fact, it was in the Teutoberg forest of Germany that the Romans suffered one of their most humiliating defeats at the height of their power, when three Roman legions were ambushed and their precious golden eagles (battle or regimental standards) were taken by German tribal warriors in A.D. 9. Part of the great difference which has remained between the culture of Latin or Romance Europe, where the Romans ruled, and that of Germanic Europe, where their influence was small, may certainly be ascribed to the forest. Gaul (as what is now France was then called), the southern part of the Low Countries, and even England (although not forested Scotland and Wales) were much more easily overrun by the Romans, even though Germany lay as near or nearer to their Italian base. In Gaul and Belgium, and to a lesser extent in southern England, agriculture had already begun to eat away at the original forests where they covered level land or blocked routes across the plains. In Germany agriculture at this period was less developed, and nomadism or hunting in the forest was more important. The hilly or mountainous character of much of Germany also tended to minimize deforestation.

In North America the forest has played a similar barrier role, as it did much earlier in originally forested south China. Forests in both places were a bar to the expansion of agriculture and settlement, an obstacle to movement and trade, and a base for dissident groups. In China the removal of this forest cover has gone to an extreme owing to the pressure of population on the land, and China proper is almost denuded of extensive stands of trees. In North America the taiga stands of Canada are still practically untouched, and even in the United

States, although most forests have been removed, cut over, or heavily invaded, it has not been necessary or profitable to do away with all of them. But the advance across the Appalachians in this country depended on the removal of the hindrances which the forest presented.

Barrier Rainforests

To find the clearest example of the barrier effects of vegetation, however, one must turn to the extreme represented by the equatorial rainforests. As pointed out above, the climate and soils of these areas are also poor, and these two factors do as much or more than the vegetation to limit human settlement. Nevertheless, the rainforest is a formidable discouragement to economic development, and the sparse populations who are able to maintain themselves within it can do so only at a very low level of development. Access, movement, and exchange are at a minimum, and regional specialization is virtually impossible. What movement does take place in the rainforest, aside from hunting and collecting, is often confined to navigable streams or rivers, the only practicable routes through such a dense wall of vegetation. Agriculture is a relatively difficult and unrewarding affair, since the massive growth of trees must first be removed, and also since the soils are generally of low fertility. Most rainforest dwellers earn their living by a mixture of shifting agriculture (see Chapter 10) and the gathering, collecting, and hunting of food in the forest: fish, roots, berries, nuts, and game. The rainforests are excellent refuge areas, and their human populations may often be composed largely of groups who have retreated or been pushed off better lands elsewhere by stronger groups. They find the protection they need in the rainforest, not only because it is hard to get at but because it is also undesirable to other people who find greater attractions in other places. The swamps and swamp forests of Georgia and Florida in the United States are also refuge areas and support small groups of technically retarded people; Walt Kelly's *Pogo* is not entirely moonshine.

The vegetation influence helps to explain what otherwise looks like a puzzle, the cultural break between Arab north Africa and Negroid Africa south of the Sahara. Here are two dramatically different civilizations juxtaposed, the Arab a relatively highly developed one, the other relatively primitive, much of it (except for the steppe border) without writing, the plough, or the wheel until these techniques were quite recently imported from outside. Usually when two such different civilizations come into contact, the technically or culturally superior one influences, dominates, or overwhelms the other. This certainly happened when the Europeans came to the New World, when Chinese and Indian civilization spread out into much of east and southeast Asia, and more recently when European pressure in the Far East has successively eaten away at the fabric of the traditional local civilizations, high though they were. But in Africa there was very little cultural or physical interchange between the Arabs and their immediate neighbors in the rainforest. The reason is not to be found in the Sahara. The Arabs, being primarily nomadic, easily and continually moved into and across the desert, and their influences on the African cultures of the steppe south of the Sahara were overpowering. In what is now northern Nigeria, for example, the Hausa and Yoruba kingdoms became Muslim in religion and were technically and politically highly developed as a result of Arab influences. The Arabs intermarried with the local Negroid people and left a mixture of Arab and Negroid peoples and cultures. But the rainforest, beyond the steppe, was a wall which the Arabs did not penetrate.

As nomadic peoples, the Arabs and their culture were fundamentally dependent on animals, mainly sheep and goats. In the desert and steppe the Arabs found a congenial area, but they were excluded from the rainforest, in part because of the tsetse fly. The tsetse carries a disease known as sleeping sickness, which is fatal to all but the smallest domestic animals (chickens, dogs, pigs); wild animals are for some reason immune, although they act as hosts for the disease. The fly likes hot, wet, dank places and is not found outside the rainforest except in water courses reaching out from it which are heavily forested, or in dense bush vegetation where there is local moisture. Obviously no people or culture dependent on domesticated animals, as the Arabs were, could maintain itself where the tsetse ruled, and the Arab contact with the rainforest was limited to occasional slave-raiding forays. The other principal economic support of the Arabs, and the most usual means of contact between cultures, was trade. Arab bases on the Mediterranean and in the Arabian peninsula, on the main sea route between Europe and the Orient, developed an early specialization in trade; the Arabs were pioneers in the sea-borne commerce between Asia and Europe and were important carriers of culture. In Africa, slaves, salt, gold, spices, and other high-value products of the steppe area and its immediate neighborhood moved north across the Sahara under Arab management. But there could be little or no trade with a primitive subsistence economy like that of the African rainforest. It produced almost no surpluses for exchange, and the physical means of exchange were severely limited. In this the rainforest played a large part. Taking all into consideration, it is not surprising that these two different cultures came into contact so little.

The rainforest is still partly responsible for a similar degree of isolation and low level of economic development in the Amazon Basin of South America and in much of southeast Asia. Most of Borneo and New Guinea, for example, are sparsely occupied by primitive subsistence groups, including, until very recent years, cannibals and head hunters, even though Borneo and New Guinea lie relatively near Java and Australia, both highly developed areas. As the vegetation map shows, the natural cover of Java was largely rainforest. But Java is small enough to be easily accessible by sea, lies in a commercially strategic location, and is fortunately supplied with fertile volcanic soil, unlike most tropical or equatorial regions. This combination of favorable factors supported a relatively dense and highly developed population which long ago removed most of the original rainforest. Along the coast of Malaya, too, where access by sea is easy, large commercial plantations have grown up at the expense of and despite the rainforest. But although local conditions, and especially access, may dent or destroy the barrier effects of the rainforest, for most of the area which it covers human activity is severely restricted.

TIMBER RESOURCES

Forest timber is a potentially valuable resource but not always an economically usable one. Most of the world's remaining forested areas, including the two most extensive types, the taiga and the tropical rainforests, are not used commercially on a large scale. Timber is a low-value commodity by weight; it cannot bear expensive transport hauls without becoming impossibly expensive in the market. Actually, the price for most timber in most markets represents mainly transport costs. Where water routes are available on which the logs can float or be carried to market, transport costs are lowest, and the major commercial logging areas of the world are on or near water. To haul logs very far by truck or train, let alone by porters or wagons, soon makes them prohibitively expensive unless they are of very high unit-value sorts. This tends to mean that forested areas far from water routes are much less heavily used and may be entirely unused for commercial timber. This is especially so when such areas also lie far from land transport lines or from lines which have the high constant volume of flow associated with low transport costs. Hence much of the taiga of Canada, Scandinavia, and Russia, where population is sparse and railways few, is not commercially usable, although the soft wood of coniferous trees is valuable especially for pulp, used to make paper.

The access and transport problems in the rainforests are even worse. As one result of the constant, heavy rainfall and high year-round temperatures, a multitude of tree species thrives in the rainforest, and the trees tend to grow in a vast jumble with no one species clustered in consistent stands. By no means all trees, or all sizes of any one species, are commercially valuable, especially where transport costs are high. In the rainforest with its special problems of access and transport, only a few species are valuable enough to bear the transport costs. Since the individual mature trees of the valuable species are scattered over a wide area, logging operations are expensive. Loggers attempt to keep down costs by cutting trees on a mass wholesale basis, and they cannot cheaply wander hundreds of yards with their equipment through a dense forest from one mahogany tree to another. Tropical hardwoods like mahogany, teak, or ebony are commercially valuable because they are dense, heavy woods and have an attractive appearance when polished. They can resist rot, cracking, sea water, and fire much better than other lighter woods. Tropical hardwoods command a very high price on the market. But the weight and density which make them valuable also make them expensive to transport. They are so dense that most of them will not float, so the rivers which are plentiful in most rainforests cannot be used to transport the timber unless the trees are first girdled and left to dry out for a year or more, or unless the logs are rafted downstream, sometimes by being lashed to lighter logs. All this takes time and costs money and helps to explain why tropical hardwoods are so expensive and so relatively little used. The great majority of the timber resources of the rainforests are commercially untouched.

In the more temperate, densely populated areas of the world where commercial logging has proceeded on a much larger basis, lumbering has been more wholesale, and huge areas have been cut off. In some cases where trees were the major local resource, the end of logging has left behind it an area of both physical and economic desolation, like a large part of the upper peninsula of Michigan. The logging industry realized long ago in Europe that selective cutting, leaving behind the smaller trees, and replanting, although more expensive in the short run, was in the long run wiser planning than a wholesale removal of all trees. This is true not only for the long-term yield of timber, but for the prevention of erosion and floods. Such practices are belatedly spreading in the United States, but not before large areas of forests have already been ruined by wasteful logging which aimed only at the short-run or immediate gain. Even under the best conditions of "tree farming," however, lumbering is necessarily an *extensive* industry, applying relatively small amounts of capital or labor to relatively large amounts of land. Regions where lumbering is an important element of

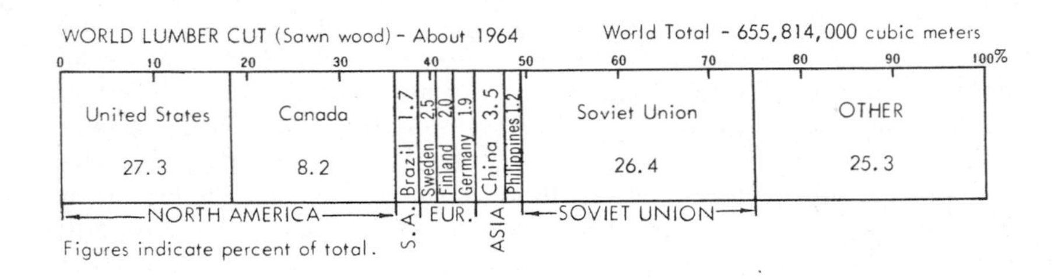

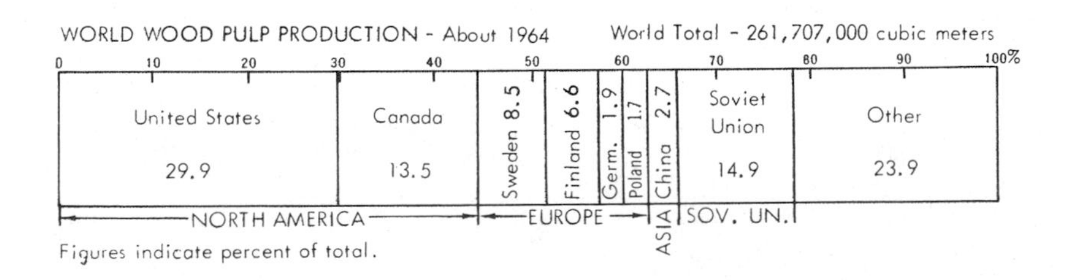

Figure 18. World Timber and Wood Pulp Production.

the economy are in part for this reason relatively sparsely populated even though they may be productive in total yield of commodities for export.

THE GRASSLANDS—ZONES OF MOVEMENT

The role of vegetation in offering an opportunity for movement is exemplified by the steppe, the great grasslands. There are no trees to act as barriers to movement, although the water problem may be serious. Most steppe grasses have very deep roots and during the long dry spells their surface leaves die off and turn brown. With the reappearance of rain, they come to life again. Trees cannot do this; they may survive dormant during the winter or during a dry season, but once their leaves and branches are killed, they die. Most trees are relatively shallow-rooted compared with most steppe grasses, and they require larger amounts of rainfall or of near-surface ground water; in general, they cannot survive outside the humid parts of the world. This can clearly be seen by comparing the world maps of climate and vegetation.

Steppe rainfall is often not enough to support agriculture without irrigation, or rainfall deficiencies may make agriculture difficult or precarious. The rainfall will, however, support grass, and the grass can support animals which in turn support man. Aside from the sedentary commercial farmers (a quite recent development in semi-arid areas, dependent on cheap world-wide transportation and on other technological and economic changes), those who live in the steppe must of necessity be in continual motion, in pursuit of grass for their animals. The American cowboy and the Argentinian gaucho, in a commercial economy, do have more or less fixed residences even though they spend much of their time moving about on the range. But in the great steppe area of Eurasia, just as in the grasslands of North and South America until a hundred years ago, most of the people have until recently been nomadic. They were not importantly dependent on markets and did not have to prepare their products for shipment, since they largely consumed what they produced. Their home was a temporary one, wherever the search for forage and water for their animals led them. Movement was relatively easy, since along the steppe or at intervals pasture and water are available. Movement is necessary because the grasses or water cannot maintain a herd in one place for long; the animals must move on before they exhaust the often sparse pasturage.

Nomadic Peoples

Nomadism is on the wane now, as the world economy and global political pressures impinge upon it. Most of the great steppe of Eurasia is politically part of the Soviet Union, and most of the rest belongs politically to China. For both countries it is highly inconvenient to have subject populations moving about with the freedom of the nomads, and the governments are also anxious to derive some economic advantage from the steppe areas which the nomads control. Over much of this area nomads are being persuaded or forced to adopt fixed residences and to produce animal products for marketing, or to take a job in a meat-processing plant. A few of them, like a group of the Kazaks, have rebelled and fled to other countries. There are still a few essentially nomadic areas left, but until a few years ago this whole vast region was a zone of continual movement. Annual migrations in search of pasturage might take a given group or tribe several hundred miles.

The nomads could also be highly mobile from a military point of view. The typical nomad doubled as a mounted tender of flocks and as a mounted warrior. He usually had a telling advantage, despite his small numbers, over the sedentary empires which lay around him, because he could strike hard without warning, quickly retire into the steppe, and if necessary strike again a hundred miles away the following day. The nomad inhabited or wandered over a poor land, and he was perennially attracted by the comparative riches and easy plunder of the better watered areas around him.

Occasionally the individualistic nomad would submerge his clannish or independent tendencies in a large organization or alliance of steppe groups and establish a military empire which swept everything before it, until in a few generations the empire dissolved into the more normal anarchy of the steppe. By such conquests, and by their usual pastoral migrations, the nomads were often carriers or transmitters of culture. In Eurasia their role in this respect was particularly important, since the steppe runs virtually unbroken from the eastern margins of Europe to the confines of the great Oriental civilizations. Contact between these two cultural and economic centers of Europe and east Asia, especially before the development of sea routes, depended on the nomads. The main route of trade between east and west followed the Eurasian steppe, and ideas flowed with trade.

One chapter in European history emphasizes the importance of the fact that the steppe penetrates well within the borders of Europe. Steppe covers most of southern European Russia (including the Ukraine) and reaches far up the valley of the Danube into central Europe. This was the highway followed by the nomadic Mongols at the time of Genghis Khan in their conquest of the largest empire in the history of the world. All of the steppe area fell under their control, after first suffering mass destruction by the Mongol armies. The center of the Russian state at that period was the city of Kiev, which lay right in the path of the Mongol advance. It was utterly destroyed and never fully regained its position. Farther north, in the forested and occasionally swampy area of European Russia, the mounted Mongols were much less effective and in the same way were also retarded in the conquest of forested and hilly south China. After the collapse of the Mongol empire, Novgorod and Moscow emerged as the new nuclei of the Russian state. In central Europe, the influence of the steppe's penetration is seen in the Asian origin of the people, language, and name of Hungary.

MAN'S MODIFICATION OF VEGETATION

Man's modification or removal of the natural vegetation may often produce more far-reaching effects than the original cover exerted. The original cover, before it is modified by man or by animals kept or encouraged by man, is the result of ages of natural selection and adaptation. This is a continuing process in which vegetation changes as climate changes or as the total pressures of animal and vegetable life change, but it can easily be upset. To remove the vegetation, or to remove one element of it, destroys what is fittingly called the balance of nature. Although this may be worth doing in terms of human benefit, it may in some cases bring results which were not anticipated and which are the reverse of beneficial. Harmful results are most pronounced in the form of erosion and floods. Vegetation, especially trees, acts like a sponge on the land. It breaks the force of rainfall, absorbs moisture, holds the soil in place, and like a reservoir, it releases moisture from its sponge of roots and soil only gradually. With the vegetation removed, a heavy shower, especially on the slope, may wash away the soil and may run off the land and into stream channels too quickly to be absorbed adequately. This contributes to worse and more frequent floods and is one of the factors responsible for troubles with the Missouri and Mississippi rivers in this country, for example, or with the Yellow River in China.

Erosion on slopes, or even on plains, will also mean that larger quantities of silt are washed down into the rivers. This may further increase floods, as well as hampering navigation and silting up irrigation works. Normal stream channels become choked with silt, and the river may change its bed. Silt deposition may raise the bed above the level of the surrounding country, as it has in the lower stretches of the Yellow and Mississippi rivers, especially where man has built dikes or levees to prevent the escape of flood waters and the silt they carry. The river itself may build up what are called natural levees, where it drops silt along its banks as flood waters retreat into the normal bed. Successive flood crests may be unable to break out, and silt will be deposited in the normal channel. With the river confined and building up its bed above ground level, the inevitable break-through of pent-up flood waters behind dikes or levees is likely to be disastrous. "Dig the bed deep and keep the banks low" is a sound prescription for river management, devised in ancient China, but it is difficult to apply where watersheds have been denuded and rivers are silt-laden.

Changes in Effective Climate

Quick runoff of rainfall may also change the *effective* climate, since the same amount of rainfall will impart less moisture to the soil. "Effective climate" refers to the usability of rainfall and the extent to which the total precipitation is made available for agriculture or for other purposes. The dust bowl of the 1930's in the Great Plains of the United States was a direct result of the plowing of the less humid parts of the steppe grasslands and the planting of wheat to meet the accumulated world demand during and after the First World War. With the soil thus exposed, a series of dry years weakened the basis of wheat growing and opened the way for winds to pick up the dry dust and bodily carry off millions of tons of soil. The drought and dust storms of Texas and adjacent areas in the 1950's are traceable to a similar set of causes, including the plowing of marginal lands after the Second World War. Steppe and arid-climate rainfall is largely in the form of convectional thundershowers, whose violence accentuates runoff and increases the vital importance of vegetation cover. Tree cover may even have the effect of increasing the proportion of milder showers, because trees decrease heat reflection and radiation from the ground. Removal of tree cover has certainly changed the *effective* climate in every part of the world where it has taken place and has also altered the soil and the rest of the vegetation cover. Animals kept by man have further reduced plant cover in treeless areas, with similar unfortunate results, or have prevented the regrowth of trees. It has been well said that "aridity is the work of man"—a pardonable exaggeration which is not as far from the truth as it might seem.

Deforestation

Marginal areas are particularly susceptible to damage, but even in better-watered places where an originally vigorous forest cover has been cut off, erosion and floods may become a tremendous problem. This is the case in much of China, for example, where extensive deforestation began some two thousand years ago. Lying at the eastern end of the Eurasian land mass, China has a modified monsoonal climate influenced strongly from the interior and characterized by a marked seasonality of rainfall. The summer peak usually comes as relatively strong showers, whereas the winters, especially in the north, may be relatively dry. This increases erosion or floods, especially when the sponge or reservoir effects of the natural vegetation have been removed. Denudation may also contribute to drought. The problem is far worse in north China and the Yellow River Basin than in the south. Rainfall in the north is more heavily concentrated in summer, its annual total is less, trees and bushes are less numerous and less vigorous, and deforestation has gone on for longer. But there have been periodic floods on the Yangtze and other south China rivers, too, aggravated by the monsoonal rainfall but due also to deforestation in the river watersheds. In a situation of milder, more evenly distributed rainfall like maritime northwest Europe at the western end of the same land mass, removal of most of the original tree cover has not been nearly as serious.

Where the original forest is marginal, it may not be able to re-establish itself by shooting or seeding once it has been cut down, even if cutting is not continued. Forests may persist for some time in areas where the climate has gradually become too dry for new forest growth but not too dry to maintain existing trees. It has been suggested that steppe or a prairie-grass type of vegetation now covers many marginal areas which were once forested, because primitive hunters or farmers burned off the original forest to drive out the game or clear the land

and set grass fires at regular intervals thereafter for the same purpose. Some students argue that much of the present steppe vegetation of the world, especially in the tropics, may be the result of man burning off an original tree cover.

Spreading of Plants

Man has also altered natural vegetation by spreading plants very widely. He has done so as long as he has lived on the earth but especially in the last four hundred years as Europeans established and intensified their contacts with the New World and with Asia and Africa. Wider distribution of useful crop plants like wheat, maize, rice, potatoes, or manioc has basically altered the landscapes, cultures, and economies of areas where these plants were absent before and of other areas related to them. Not only useful or cultivated plants have been spread but many others, usually as "volunteers." In few parts of the world does the present supposed "natural" vegetation not include plants introduced purposely or accidentally by man, often from another continent. Hybridization, natural and human-directed, further complicates the mixtures. Man's modification of vegetation by fire, clearing, plowing, road building, drainage, or other activities also favors some plants (as it favors some animals, such as mice, deer, pheasants, or starlings) at the expense of others. This may fundamentally change the total vegetation complex, apart from the cultivated plants or those which man purposely tries to encourage or eliminate. Perhaps the simplest illustration is the profusion of weeds in a plowed field, many of which may be of distant origin, and all of which have multiplied because of man's activities. Recent illustrations of the effects of grazing animals on vegetation, apart from the well-known ravages of sheep and goats, were provided by the introduction of the rabbit into Australia, and its near extinction in Britain as a result of a pandemic disease. In both cases the vegetation and the look of the whole landscape were changed.

But society's efforts have been concentrated on the removal of vegetation, especially the forest. Man has had an antipathy for the forest ever since the emergence of *Homo sapiens*, and probably before. The forest's dark recesses harbored wild beasts and other unknown terrors, and from a purely practical point of view it was in the way of agriculture and of circulation. Modern man has only recently, after many unfortunate experiences, learned that in some cases retention or careful modification of the natural vegetation may be better for him in the long run than its removal.

Questions for Further Study and Discussion

1. Why and how should Paleolithic and Neolithic man have modified or reduced the forest?

2. Where are the remaining major forest stands in the world? Which are currently being exploited commercially for lumber on a large scale? Why in each case? Why has each of the other forest areas been less used?

3. What other regional examples can one find of the impact of nomadic groups on sedentary civilizations?

4. What major areas of the world are being reforested? Why? Is it for the same reasons in each case?

5. Why does such a large share of the world's marketed tropical hardwoods come from Burma and Thailand? How are the logs felled and transported from forest to market? Why? Why do not more tropical hardwoods come from the rainforests of Africa and South America? Why does so much come from the Philippines?

6. How was natural vegetation related to the regional differences in pre-Columbian Indian cultures and economies in North America?

7. Why is vegetation such an important element in all river-basin planning programs?

8. The Food and Agriculture Organization of the United Nations estimates that about 10 per cent of the land surface of the world is cultivated in crops. Why so little? Is it likely to increase? Why or why not?

9. Can you find place names in the United States which suggest that forest once covered the sites? Can you find any American parallel to Robin Hood?

10. How in detail did forests affect each stage of the westward movement in the United States? What role did they play in the so-called French and Indian wars from 1689 to 1763?

11. What plant and animal species have multiplied as a result of man's activities? Why in each case?

Selected Samples for Further Reading

Anderson, E. "Man as a Maker of New Plants and New Plant Communities." In W. L. Thomas, ed., *Man's Role in Changing the Face of the Earth*, pp. 763–77. Self-explanatory.

———. *Plants, Man, and Life*. Boston, 1952. A clearly written, semi-popular discussion of man's dependence on plants and the variety of ways in which he has used and abused them.

Bacon, E. E. "Types of Pastoral Nomadism in Central and Southwest Asia," *Southwestern Journal of Anthropology*, X (1954), pp. 124–36. The study of an important adjustment to the environment of the great Eurasian grasslands, and some of its implications and consequences.

Clark, J. G. D. "Farmers and Forests in Neolithic Europe," *Antiquity*, XIX (1945), 57–71.

Colman, E. A. *Vegetation and Watershed Management*. New York, 1953. A broad survey of the basic role played by vegetation in restricting runoff, preventing floods, and conserving water supplies.

Curtis, J. T. "The Modification of Mid-Latitude Grasslands and Forests by Man." In W. L. Thomas, ed., *op. cit.*, pp. 721–36. Self-explanatory.

Darby, H. C. "The Clearing of the English Woodlands," *Geography*, XXXVI (1951), 71–83, and "The Clearing of the Woodland in Europe." In W. L. Thomas, ed., *op. cit.*, pp. 183–216. Two studies in historical geography.

Eyre, S. R. *Vegetation and Soils: A World Picture*. 2nd ed. London, 1968. A clear general treatment, laying special stress on the role of vegetation in soil formation, with appropriate attention to the agency of man.

Guest, S. H., ed. *A World Geography of Forest Resources*. New York, 1956. A collection of articles by various authorities, considering the relation between man and forests, the use of timber as a resource, and regional summaries from a variety of areas.

Harris, D. R. "New Light on Plant Domestication and the Origins of Agriculture," *The Geographical Review*, LVII (1967), 90–107. A detailed review of recent literature on this topic.

Holscher, C. E., and Spencer, D. A. "Sheep, Goats, and Grasslands." In United States Department of Agriculture, *Grass*, 1948 Yearbook of Agriculture, pp. 94–98. Washington, 1948. The effect of close-cropping animals on vegetation. The volume contains numerous other valuable articles on grass and the grasslands.

Huzayyin, S. "Changes in Climate, Vegetation, and Human Adjustment in the Sahara-Arabian Belt, with Special Reference to Africa." In W. L. Thomas, ed., *op. cit.*, pp. 304–23.

Küchler, A. W. "Classification and Purpose in Vegetation Maps," *Geographical Review*, XLVI (1956), 155–67. An outline of some of the many problems of mapping vegetation, including three sample maps.

Malin, J. C. "Man, the State of Nature, and Climax, As Illustrated by Some Problems of the North American Grassland," *Scientific Monthly,* LXXIV (1952), 29–37.

Meyer, H. A., et al. *Forest Management.* New York, 1952. A textbook of forestry practices.

Richards, P. W. *The Tropical Rainforest: An Ecological Study.* London, 1957. Somewhat technical, but including useful discussion of classification types and a detailed treatment of the selva environment.

Richter, C. *The Trees.* New York, 1940. A novel of the advancing frontier across the Ohio River when the land was an unbroken sea of trees—a vivid picture of man's struggle with the forest.

Sauer, C. *Agricultural Origins and Dispersals.* New York, 1952. Theories of the location, nature, and development of the earliest agriculture, the spread of plants and cultivation methods, and the later development of pastoralism.

Stewart, O. C. "Burning and Natural Vegetation in the United States," *Geographical Review,* XLI (1951), 317–20.

United States Forest Service. *Timber Resources for America's Future.* Washington, 1958. A collection of articles by various authorities, including consideration of the effects of expected population increases.

Young, A., and Riley, D. *World Vegetation.* Cambridge, England, 1966. A paperback booklet with fine photographs illustrating the variety of vegetation types and including a general text which provides a good survey and relates vegetation groups and species to climate and to soils.

10
Soils and Man

AGRICULTURE IS STILL the most indispensable productive activity, and soils are therefore a primary resource, more important than coal or iron ore or petroleum. Like mineral deposits, soil can be exhausted, although broadly speaking soils may be a product of climate in a more direct or current sense than is the case with coal or petroleum. It still takes thousands of years, however, even if not the millions needed for coal, for a maturely developed soil to form. Since on such a time scale climate may change considerably, the pattern of soils and their distribution is complex. Most present soils are probably in large part "fossil" soils derived from an earlier glacial or preglacial period and also acted upon by subsequent climates. There are factors other than climate which influence soils, which are discussed below. Nevertheless, the general outlines of the major soil groups of the world coincide roughly with climates in their distribution, as can be seen by comparing the two world maps. It was the Russians who first realized in systematic fashion a causal relationship between climates and soils; a careful study of the two maps will show that the Russians had certain advantages in developing this idea.

A number of new soil classification systems have been developed as the result of research during the past couple of decades which attempt especially to reflect soil formation, like all other environmental elements, as a dynamic process and which also allow for the great range of factors which affect this process. Soil classification systems are complex, and involve technicalities and detail which need not concern us here, since our focus is on man's *use* (and misuse) of his physical world. However, it should be recognized that the discussion of soil types here is highly general and that a great deal more information is readily available, increasingly now from the work of the Soil Survey Staff of the Soil Conservation Service of the U.S. Government. That agency has recently published a new classification system titled (appropriately enough) "The Seventh Approximation." A good clear discussion of this system, and of the technical properties of soils and their classification, can be found in J. E. Van Riper's *Man's Physical World* (see the reading list at the end of this chapter).

SOILS AND CLIMATE

Climate may allow or prevent the retention or accumulation in the soil of soluble mineral elements and of *humus* (decayed organic matter). Most minerals in the soil are soluble, except silica, iron, and aluminum oxides; where rainfall is heavy and continuous and temperatures are high so that oxidation is rapid, most humus and minerals except silica, iron, and aluminum oxides are removed. *Leaching* (removal of soluble elements by water) and oxidation, when carried to extremes, produce a material which is not only infertile but may be literally as hard as a rock. Such soils are frequently called *laterites*, from the Latin word *later*, meaning brick, because they may be as hard as a brick and are often given a brick-red color by insoluble oxidized iron. By no means all red soils are laterites, nor are all laterites red, since redness is simply the result of the presence of oxidized iron, which occurs in a great many soils in warmer climates where oxidation is relatively rapid. Laterites are especially thoroughly leached and oxidized. In many parts of the tropics where laterites occur they are actually quarried and used as building or paving stones. Leaching and accumulation of iron and aluminum also occurs in the grey, acid soils of humid but cooler climates, where such soils are frequently labelled *podzols*. It has recently been demonstrated that podzolization and laterization are in effect the same process, resulting in essentially the same end product, even though the color, texture, and compactness of the soils may differ as a result of lower oxidation in cooler climates. (See the reading list for this chapter.)

Leached soils are generally of low fertility and are particularly deficient in the highly soluble mineral nutrients which help to produce protein in the crop. Such soils may be able to produce good yields of bulk carbohydrates, but protein content will be low, and the crop will therefore be less valuable as food or as animal feed. This is a common problem in the wet tropics and has also been analyzed in some detail by comparing humid and arid areas in the United States (see the reading list for this chapter). The luxuriant forests which occur on many tropical lateritic and cold-climate podzolic soils, benefiting from plentiful moisture or high constant temperatures, give a misleading impression of soil fertility. Such soils cannot usually support annual field crops for more than a short time before their slight fertility is exhausted, since most annual crops are more soil-demanding than trees.

The effectiveness of a given amount of precipitation depends in part of the evaporation rate, as discussed in Chapter 8. Ten inches of rainfall in northern Canada with low evaporation may produce a highly leached soil, whereas ten inches in southwest United States with high evaporation will not. Where the effective rainfall is slight, more of the soluble mineral elements are retained and the soil is likely to be much more fertile. High constant temperatures also speed up the decomposition of humus, or dead vegetation, and its complete oxidation and leaching. Low temperatures retard decomposition and oxidation and may lead to acid soils or soils poor in humus content, since the humus merely lies around as surface litter and generates acids. Podzolic soils are formed under these conditions, whereas laterites are formed where high constant temperatures allow little or no accumulation of humus. True laterites are found only in the equatorial climates, and true podzols only in the humid, cold climates. Between these extremes and the desert regions there is a complex gradation of soil types which broadly reflects the accompanying gradation of climates. As usual, the extremes are unfavorable: in the equatorial tropics, laterites, in the cold areas, podzols. In the desert, the deficiency of rain may mean the absence of humus and the accumulation of salts or alkalis which are not washed away as they are from the soils in humid climates and which poison the soil for plants.

Humus

Vegetation plays a major role in the development of soils, by providing humus and by protecting the surface layers. Humus is probably the most important positive factor in soil building. The most fertile of all soils lie in the general zone of the steppe, where decaying humus from deep-reaching grass roots has built up a deep, dark, rich layer of earth. Humus is important for the structural qualities which it gives to a soil, promoting good circulation and retention of air and water, and for the rich plant nutrients it provides. Humus is also an essential aid in the process of biological and bacterial action which helps to keep a soil aerated. Good soil is alive in the sense that it supports a myriad of bacteria, earthworms, and other creatures. Their work of decomposition and circulation and their own dead bodies are of tremendous benefit to plants. This live matter is often present in a soil in direct proportion to the amount of humus which the soil contains. Where vegetation is absent or sparse, or where vegetation decomposes too rapidly or too slowly, the productivity of the soil suffers seriously, apart from the soil erosion which may also accompany the absence of vegetation. Some types of humus are better for most crop plants than other types. For example, coniferous or needle-bearing trees yield a humus which is inclined to be acid; where such trees are the principal suppliers of humus, only plants which like or tolerate an acid soil will thrive. Such plants include the conifers themselves, the rhododendron family, pin oaks, blueberries, hollies, and some lilies, but very few of the major commercial or food crops.

Parent Rock

What is called the "parent material" also affects the nature of the soil. The soil's original content of minerals is inherited from the parent rock, and soil structure or texture may also be derived from the parent materials. Heavy clay soils may be the result of weathering of one type of rock, sandy soils of another. Soils formed, for example, in regions of limestone rock are often likely to be fertile, those in regions of granitic rock less so. But climate exercises an over-all influence, leaching away the fertility of an originally well-provided soil if it happens to lie in an area of heavy rainfall, or allowing the accumulation of humus and retention of minerals in a granitic soil if it lies where rainfall is moderate or slight. Small areas in every part of the world reflect in their soils the parent rock. Europe and the United States, for example, are dotted with fertile limestone-soil areas which contrast sharply with surrounding less fertile areas and with regions of heavy clay soils where poor drainage affects natural vegetation and cultivated crops. In most of Europe and America rainfall and temperatures are moderate enough to allow fertility to be maintained, at least to a degree and for a time. But even in the rainy equatorial tropics local soil distinctions based on parent rock are apparent.

A somewhat special case is soils formed from the weathering of volcanic lava or ash deposits. Such soils are often deep enough to resist leaching, and they may be frequently renewed by periodic lava or ash deposition, so that climate affects their fertility relatively little. Where their chemical composition is basic rather than acidic, they may be very productive agriculturally. The island of Java has already been cited in Chapter 9 as an example, and there are many other similar areas of fertile volcanic soil, such as parts of Cuba, the Columbia plateau in northwestern United States, and parts of the Deccan plateau in south India. In broader patterns, however, the role of climate is clear. From the same original materials in two different climatic situations, two quite different soils will tend to develop.

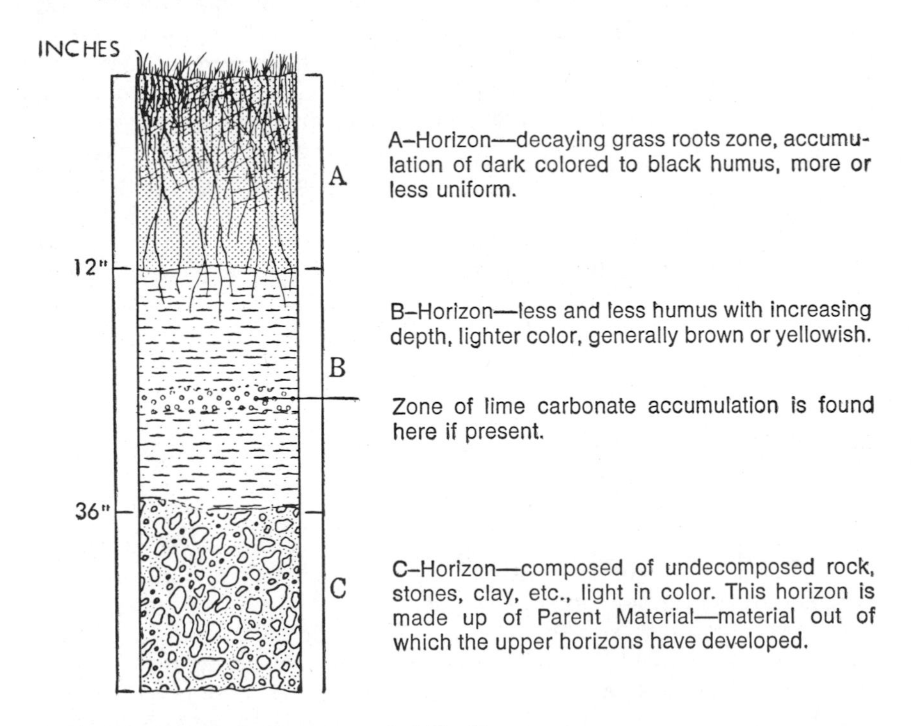

Figure 19. Typical Chernozem Soil Profile.

Transported Soils

Some of the most productive soils are the result of erosion and subsequent redeposition in another area, by water, wind, or ice. The materials which rivers transport and redeposit along their courses as *alluvium* stand out on the world map of soils as they should, for they form among the most fertile of all soils. The river as it flows through a variety of rock and soil areas picks up a great mixture of sediments. River silt is thus likely to be well balanced chemically and to contain most of the mineral elements necessary for plant growth. The usually deep soil formed by river deposition is also likely to have good drainage and structure and a fine texture, depending on the velocity of the river, for a stream can carry larger particles or even rocks in proportion to the force of its flow. As the velocity decreases, the larger particles are dropped, and the finest alluvium is found in the lower courses of rivers or in their deltas where, after reaching a relatively flat plain, the rivers are flowing more slowly. Areas of alluvial soil, as these riverine deposits are called, are particularly valuable when they occur where neighboring soils are poor, as in much of the tropics. This is an important reason for the remarkable concentration of the population of India, China, and peninsular southeast Asia in river valleys. But alluvium nearly everywhere in the world is at a premium. *Loess* is akin to alluvium, but is believed to have been laid primarily by wind rather than by water. Strong outblowing winds from the retreating glaciers at the end of the last ice age may have picked up finer particles from the glacial outwash plain, deposited earlier by melt water from the ice, and distributed this dust in thick layers on top of extensive areas to windward. Loess areas include much of north and northwest China, parts of northern Europe, and parts of central United States. In fertility and structure, loess is similar to river alluvium. Many large areas are also covered with soil originally dumped there by advancing or retreating glaciers, or by the gradual shrinking of lakes formed by the melting of ice sheets. Some of the glacial deposition may be stones, gravel, or sand, as in the case of materials deposited by swift-flowing rivers, but river alluvium may also be matched by fine-textured and well-balanced glacial deposition of varied origin which is highly productive agriculturally, for example, large areas of the northern half of the American Middle West.

WHAT MAKES A SOIL PRODUCTIVE

There is no need to be concerned here with the technical aspects of soils and soil types. Actually, even the soils scientists are increasingly reluctant to distinguish general soil types on a world-wide basis and are not sure how important color, texture, levels, and other distinguishing marks are to the agricultural properties of a soil. Soils are of interest here only as they may help in understanding man and his works. The major question is whether a given soil is actually or potentially productive, or whether it may have special characteristics which fit or unfit it for specific uses. Fertility is one of these characteristics and one measure of a soil's productivity. Fertility in its strict sense means the available mineral content of a soil and in particular its content of adequate soluble amounts in the right proportions of the main elements essential to plant growth: nitrogen, phosphorus, potassium, calcium, and others in lesser amounts. The structure, texture, and depth of a soil are also of great importance, since plants depend on a circulation of air, water, and nutrients in solution to their roots. A densely packed soil or one with either poor drainage or with so much drainage that the water percolates right out of it will obviously not have proper circulation. However fertile it may be in its mineral components, it will not be very productive. Soil which is too shallow also hampers proper plant growth; it quickly dries out, and drainage is likely to be poor. Some soils can be plowed more easily and effectively than others. Some retain good circulation of air and water after plowing, while others cake together with the first rainstorm or bake hard with the first dry spell.

As a rule, lack of fertility is easier and cheaper to remedy than unfavorable structure, texture, or depth. Chemical or organic fertilizers can be applied, but the technical means of correcting soil structures are more limited. In both cases, however, it is economic rather than technical considerations which restrict the artificial improvement of a soil. It costs money to fertilize, to dig in humus, to lay drains, or to apply the recently developed chemical soil conditioners. It pays to spend this money only when an appropriate return can be realized. In practice, such improvements are generally limited to the favorable climatic areas, to irrigated areas, or to areas characterized by intensive land use because of population pressure or because of market considerations. It often may not pay to fertilize, for example, the huge wheat fields of semi-arid, thinly settled, and relatively remote western United States, Canada, or Australia. In their marginal climates the cost of fertilization would not be repaid by profitable increases in wheat yields. It does pay to fertilize, and if necessary to lay drains, in the irrigated Imperial Valley of California, the truck-garden areas of New Jersey, or the rice-growing deltas of the Orient. In each case location or local conditions are such that soil improvements are worth while.

MAN'S TREATMENT OF THE SOIL

Fallowing and Rotation

Man modifies the soil in many ways. Fertilization with chemicals, pulverized rock, manure (which is rich in nitrogen), or humus is only one way, and not necessarily the most important. Depletion of the soil may be slowed by *fallowing*, or allowing the soil to rest for a shorter or longer period between crops. Where the soil is naturally not very fertile or where the crop being grown is especially soil-exhausting, fallowing is often essential. It is characteristic of the agriculture carried on in large areas of the tropics outside the alluvial or volcanic soil regions, where heavy rainfall aided by strong relief and by high constant temperatures has badly leached the soil. Farming in these areas is often *shifting agriculture*. A given plot may be cultivated for only one, two, or three years before its soil is exhausted and the farmer moves on to clear another plot. He will return to the first plot only after several years of fallow, the period varying with the type of soil, climate, relief, and crops grown, and also with the pressure of population. On each small plot in turn, the forest is cleared by burning, girdling, or felling, and the litter is burned. The ash provides a quick fertilizer. When the plot is abandoned, the second-growth jungle in a tropical climate quickly recovers the land so that it is protected from erosion. Shifting cultivation is often called a wasteful form of land use. It progressively or continually removes the natural forest cover or second growth, and any one farmer must range over a very wide area, from which he can produce proportionately little. But it is difficult to devise a less wasteful or more productive agricultural technique for such areas, especially since the technical or economic advantages of a commercial economy are largely lacking. Fallowing is practiced in most agricultural systems to some degree, and in drier areas it also has the advantage of conserving the moisture in the soil, which is of course drawn on heavily by any crop. In the European Middle Ages, fertilization was not well understood or heavily used, and fallow occupied one third of the "three field system" in order to restrict soil exhaustion.

Another widely practiced method of conserving soil fertility is the growing of crops in *rotation*. Each crop takes a different combination of nutrients from the soil. If the same crop is grown in the same field repeatedly, yields tend to decline. But if the same field can grow a variety of crops in successive years, the yields of each can be better maintained and the soil balance better preserved. Some crops actually add nutrients to the soil, most importantly the *legumes*, which include peas, beans, clover, peanuts, lentils, and alfalfa. Leguminous plants extract nitrogen from the air, which then forms in little globules on their roots. They leave a soil more fertile, especially for heavy nitrogen-demanding crops like most cereal grains, than it was before the legumes were planted. Legumes are therefore often used in crop rotations, alternating with grains.

Other crop plants may help to aerate the soil by their root development, especially if they are uprooted when harvested, sugar beets, for example. Still others help to hold the soil in place and protect it against erosion by water or wind, whereas some are not good *cover crops* and invite erosion, especially on slopes. Most row crops are not good cover crops, since they leave much soil exposed between the rows. Cotton, maize, and tobacco are particularly conducive to erosion, both lateral and vertical, since rain percolates through exposed soil more rapidly. Cotton and tobacco also make especially heavy demands on soil nutrients. Since each crop has its own particular soil demand or soil preferences, an intelligent farmer, if he has adequate choice, must use care in combining and rotating crops and soils, especially if he wants to preserve his soils, as well as maximize his yields. Most farms grow more than one crop, in part for this reason, since even within a small area soils, drainage and soil moisture, slope, and often even isolation and air drainage (important for minimizing frost damage) will vary. The growing of different crops also helps to spread labor needs more evenly over the agricultural year, restricts the spread of plant diseases and pests, and minimizes financial risks due to failure of any one crop or to varying market prices.

Cultivation and Weeding

Man also alters the soil by cultivating it. Strictly speaking, cultivation means the stirring or aerating of soils by plowing, digging, scratching, raking, or sometimes just by making holes in it with a pointed stick or other instrument. All these techniques help air to circulate in the soil and also break the soil up into finer particles so that water and nutrients in solution can circulate better. Cultivation may also improve drainage, at least in the upper layers. Although it may increase evaporation of water from the soil, cultivation can also help to reduce surface evaporation if it pulverizes the surface layer finely enough and often enough to break up the capillary action by which water moves to the surface. Frequent harrowing, or pulverizing of the surface layers, is commonly done in dry areas to conserve soil moisture, especially during fallow periods when an additional surface layer of humus, or even of crushed stones or pebbles in areas of sparse vegetation, may be applied as a *mulch* to restrict evaporation.

Plowing can accelerate erosion, especially if the furrows run in the same direction as the slope, or can significantly decrease erosion if the furrows follow the contour of the land at right angles to the slope. *Contour plowing* has saved many farming areas from destruction and can help to safeguard almost any cultivated land. Even a slight slope hardly perceptible to the casual eye may generate serious erosion if it is left bare in a climate of heavy rains, and on perfectly flat land heavy rains may produce *splash erosion*, especially if the vegetation cover is inadequate. Some soils, the lighter or sandier ones, erode more easily than other heavier or clayey soils.

Most soils in areas of moderate or light rainfall profit from frequent cultivation. Usually, the deeper the plowing, the better the soil retains its productivity, not only because aeration and circulation are improved, but because soil layers beyond the reach of most annual plant roots are turned over by a deep plow and brought to the surface or to the root zone where plants can use their less exhausted nutrients; this depends, of course, on how far below the surface the soil remains fertile. The soil also profits if some of the annual crop can be left behind after harvest—roots, stalks, leaves, or whatever is not going to be consumed—to form humus and to replace some of the minerals which the crop has taken from the soil. Agriculture breaks the natural cycle of plant growth and decay by removing the plant, or part of it, when it is mature so that its decayed materials are not returned to the soil. Unless the nutrients can somehow be replaced, the soil suffers. Agriculture also interferes with the natural plant and soil balance by allowing only one crop plant in any one field instead of the natural profusion of plant communities and thus puts a heavy strain on the soil. Weeds take nutrients and water from the soil, too, and leave proportionately less for the desired crop. If weeds can be left in the field to rot when they are cut or pulled up, the soil also profits and, with the dead weeds as a mulch, the soil will have extra surface protection against evaporation.

These generalizations about cultivation and weeding apply well enough to soils and agriculture outside the tropics. But where rainfall is high in amount and violent in character, or where temperatures are constantly high, leaching and erosion are accelerated by the clean weeding of a field and by the plowing of the soil. Traditional farming practices in many parts of the wet tropics try to leave as much vegetation cover as possible on the soil, to weed as little as possible, to avoid row crops, and to plow as little as possible or to rely on aerating the soil with a pointed stick. Crops may be grown crowded or jumbled together, and the field may look most disorderly and poorly cared for to an outsider. It is only recently that technical advisers from outside the tropics have begun to realize that such practices make good sense.

Agriculture and Population Pressure

Population pressure exercises a powerful influence on land use, and indeed one can analyze all agricultural systems and distinguish between them in terms of population pressure. Where few cultivable acres must support many people, as for example in much of monsoon Asia, most farmers cannot afford to leave the roots or stalks of their crops in the field; crop residues are needed for fuel or for animal fodder. Few Asian farmers can afford to apply adequate amounts of manure, for the agricultural system cannot support enough animals, and manure is also in demand as fuel. Few can afford to fallow their fields or to rotate crops; since the farmers must obtain maximum returns per acre every year, they must devote their land to growing the one most productive crop year after year or season after season.

Fortunately it is rice, grown for the most part in standing water, which dominates the agricultural system of much of monsoon Asia, and rice is in many ways a unique crop. It takes nutrients from the soil, but they are largely if not entirely replaced by nutrients which enter with the irrigation water or are derived from animal and vegetable life in the irrigated paddies, especially the blue-green algae which thrive in the high temperatures and which are potent fixers of nitrogen from the air. The water in the paddies also protects the soil from the damaging effects of wind and rainfall erosion and leaching as well as from evaporation. Other nutrients are added by the farmer, including night soil (human manure). Weeding and cultivating are, however, assiduously practiced, methods are in general highly *intensive* (much labor applied to relatively small amounts of land), and crop yields per acre are very high. In the less densely populated parts of the world fewer people have to be fed from each cultivated acre, and farming can be more *extensive*. Where income from manufacturing and trade is relatively high and agriculture is not the only economic support, good farming practice often plows under a crop of legumes or of something else in order to enrich the soil without taking away any of the crop. This is called *green manuring*, and it is fine for the soil, but clearly the peasant farmer in monsoon Asia cannot afford to do it. Population pressure also helps to explain shifting cultivation and the changeover to sedentary agriculture when it occurs, irrespective of soils. As population mounts in any area, agriculture must become more intensive and produce higher per-acre yields.

Man's Effect on the Soil

Whether he improves the soil or degrades it, man alters it whenever he uses it for any purpose. Chapter 9 has already examined some of the effects on the soil which modification of the natural vegetation may have. If only out of immediate self-interest, most farming tries to improve the soil, or at least to make it remain productive. Good farming methods can transform an unproductive soil into a highly productive one, if the farmer has the technical means and the economic ability to invest labor and capital in the soil. In balance, most long-established farming is of this positive and constructive sort rather than the destructive farming which is often called "soil mining," because it takes from the soil until fertility is exhausted and does not contribute proportionately to the land. Such practices, apart from shifting agriculture, may mean that the land will have to be abandoned after a brief period of heavy use. Too much of American agriculture has unfortunately been of this "soil-mining" type. Cotton and tobacco, maize and wheat, all important in the United States, are soil-exhausting crops which have often been grown for short-term gain. Application of chemical fertilizers alone seldom restores the soil balance adequately. The agricultural system in current use may often be the major factor in determining what the soil can physically support or what the soil's physical properties are, since all farming transforms soils, for good or bad.

Tree Crops

Perennial or tree crops are usually not as sensitive to soil conditions as are annuals, which must in effect produce an entire plant from seed in one growing season on the basis of what the soil provides. Most tree or perennial bush crops do need good drainage, and since their roots go deeper than most annuals, orchards and plantations must often pick sites and soils with some care, avoiding those which are poorly drained. Fertility may be less important, and poor soil areas may be well forested or may support good tree crops. Orchards often make use of slopes, where drainage is likely to be good but where soils are often poorer than on level land because they are more subject to erosion and to soil slip and slide. Trees help to hold the soil on slopes, and fruit trees may benefit on slopes by avoiding the lower pockets of land where cold air tends to collect and frost damage is higher; fruit blossoms are especially susceptible to frost damage. In the tropics, where most plantation agriculture is located and where most soils tend to be poor, tree crops like rubber, tea, oil palm, coconuts, or cacao (chocolate) can yield well on lateritic soils which could not profitably support annuals. Trees do, however, reflect soils, like all vegetation. Outside the tropics, coniferous trees are usually more tolerant of poor soils (and of cold climates) than are the hardwood or deciduous trees.

ECONOMIC AND SOCIAL INFLUENCE OF SOILS

Soils may often be reflected in the social or economic structure of an area. A good soil area may be a relatively prosperous one, not only for agriculture but for the many service industries, market centers, and large population which a rich farming area may support. The corn belt in the United States is a good example. The productivity of corn-belt soils, in this warm-summer, moist climate, is responsible in large part for the size and manifold functions of the city of Chicago, which is primarily a service, trade, and manufacturing center for the agricultural Middle West. Corn-belt soils and climate play their part also in the size of New York City, and indeed of the whole national economy, which in differing degrees finds markets and services to perform in or for the populous and prosperous corn belt. As a primary resource, soils are an important item of national strength. They may permit considerable self-sufficiency or surpluses for export in agricultural products, or they may prevent surpluses and help to require expensive or strategically awkward imports. Regional or national soil differences, like other regional differences in environment, may help to stimulate complementary exchange between regions or nations.

A good soil may also attract heavy investment in the form of irrigation or in the form of specialized plantation or truck-garden agriculture. A poor soil area is likely to be much less prosperous as a whole, unless there are compensating advantages in other respects, and its low economic status may be reflected in the size and condition of its houses, its roads, its schools, and perhaps its political sympathies. In middle-western United States, where some study has been made of the correlation between soil types and voting habits, there has been considerable coincidence between conservative politics and good soils on the one hand, and between liberal or radical politics and poor soils on the other. (See the reading list for this chapter.) If soils are very good, they may be reflected in a high price for land and may therefore in some cases be cultivated largely by tenants who cannot afford to buy the land but who pay a high rent for it. The land may be very productive without returning a high average income to the individual cultivators, or it may support a reasonably prosperous tenantry. A map showing tenancy in western Europe would show a clear relationship between high tenancy and the combination of fertile soil and favorable climate when the spatial distribution of tenancy is compared with the spatial distribution of soils and climate.

In other cases good soils may attract a dense population into a semi-arid or fluctuating climate, as in north China, where the shortage of rainfall contributes to the fertility of the soils. Recurrent famines often resulted because there were limited means of regional exchange to supply the area with food when rainfall was inadequate and crops failed. Good soil may sometimes join with favorable climate to invite what may be called "fussy" crops, which are not tolerant of low fertility or poor soil structure and are climatically demanding. Cotton, tobacco, sugar cane, indigo, and jute are fussy in this sense. Land which can grow them is likely to be expensive, and these crops also require large amounts of hand labor (in some cases recently displaced to some extent by machinery). In many areas where these crops have been grown, slavery, tenancy, or contract labor has been the solution applied.

Soils and the American Civil War

Maps of soils and climate both show a break in their patterns for eastern United States which coincides almost exactly with the line between Confederate and Union territory or sympathies during the Civil War. What is now West Virginia, still called by some of its inhabitants "West by God Virginia," lies outside the soil and climate patterns which cover most of the Old South. As a largely mountain or upland area, West Virginia's relatively poor soils and cooler climate could not support the plantation agriculture worked by slaves in lowland Virginia and in most of the rest of the Confederate south. South of West Virginia the Appalachians have these relatively poor upland soils, and there, too, in the mountains of Tennessee and the Carolinas, Confederate sympathies were often lacking. Most of Kentucky, the poor-soil upland parts of it, belonged in the northern camp politically. One of the strongholds of plantation agriculture in the south was and is the alluvial-floored valley of the Mississippi, a logical center for the institution of slavery. Slavery was the root and body of the conflict between north and south, and it was clearly related to these broad differences in soil and climate between the two combatants, although there were, needless to say, many other factors than soil and climate differences which accounted for slavery. Similar distinctions between types of social and economic organization, though seldom with the same tragic consequences, can be made in part on the basis of soil and climate differences in many other parts of the world.

Soils and Urban Sites

In special cases, local soils may affect urban siting, since this involves questions of drainage, foundation for buildings, and vulnerability to earthquakes. Recently deposited alluvium may be avoided for large buildings in areas where earthquakes are frequent. Uneven settling and foundation problems in some soils, especially alluvium, may make special and expensive building techniques necessary. Soils with poor drainage pose problems for sewage, roads, and the maintenance of structures. Usually, such local site problems are less important than wider considerations of location and access. In fact, many cities have arisen on poor local sites because there was no preferable alternative without sacrificing access. All of the site problems mentioned above, and many others, are common to most river deltas, for example, and yet many of the world's largest cities have grown up on deltaic alluvium. Location near the mouth of a navigable river is advantageous enough to outweigh problems of drainage or building. New Orleans is a classic example; much of the city actually lies below the level of the Mississippi River, which at New Orleans is confined behind natural and artificial levees and has built up its bed by silting. One of the problems of New Orleans is sewage disposal, since wastes cannot be dispersed by gravity and have to be pumped up to the river. There are many other deltaic cities in other parts of the world which have grown despite local soil or other site problems. Freetown in Sierra Leone has a different sort of problem. The vicinity of the city is covered by a particularly hard and impermeable laterite crust which has discouraged even natural vegetation except for some sparse grasses. Freetown has grown up in a small "island" where there is a surface layer of workable soil over the laterite crust.

Intensive and Extensive Agriculture

Soils play a part in determining whether an agricultural system will be intensive or extensive. Intensive cultivation is more likely to be profitable on a good soil than on a less good one, if it is combined with a favorable climate or if it can be irrigated. A poor soil, or a soil in a less favorable climate, is more likely to be cultivated extensively since it is less likely to return an adequate recompense or yield to repay intensive investment. There are many exceptions to this. The fine steppe soils are for the most part extensively cultivated, and the much poorer soils of densely settled parts of the tropics or subtropics are worked intensively. Climate is largely responsible for these exceptions, but it does not entirely destroy the association of soils with the degree of agricultural intensiveness. Poor soils are worked intensively, as noted above, on tropical plantations which specialize in tree crops. The availability of capital and the pressure of population are clearly other factors which will affect the intensiveness of soil use. But soils are one factor which helps to determine whether investments in any form will pay.

WHAT DETERMINES SOIL USE

The uses to which a given soil may be put will depend for the most part on the following factors:

What the Soil Will Physically Support

The soil demands or soil tolerances of different crops vary widely. Fussy crops have particularly stringent requirements. Others are tolerant of poor soils, or of soils with particular characteristics. These less demanding crops would include, in addition to trees, potatoes, rye, millet, and hay. Where they are grown they may provide some evidence of poor soils, although climate may be even more important in accounting for their presence. Dairying often tends to occupy poorer soils because it does not need good ones (for pasture) as much as field crops need good soils and is therefore less willing to pay for them. Poultry raising is often in the same position.

The Nearness, Nature, and Size of the Market

Poultry raising and dairying are especially sensitive to market location, because their products are perishable, bulky, or present proportionately expensive transport problems. But the best soil in the world is of little use for any crop without an economically accessible outlet for its products. The location pattern in the dairy industry of cheese, butter, and fluid milk production at different distances from the market has already been discussed in Chapter 3. Most large cities, at least in Europe and North America where milk is important in the diet, support fluid milk production immediately around them, whether the soil and climate are suitable for dairying or not and even if fodder must be brought in from other areas. Poultry, especially eggs, are also dependent on quick access to market, and the industry tends to cluster on poorer soils in the vicinity of large cities, sometimes even in their suburbs. Generally, the farther a soil is from the market, the more extensively it will be used, although it may often be devoted to high-unit-value crops or to crops which are easily and cheaply shipped long distances. Wheat is a good example. It keeps well, ships easily, and can be loaded and unloaded mechanically. Most commercially grown wheat comes from the steppe areas of the world, which are remote from major population centers. Wheat is more tolerant of dry climates (though not of poor soils) than most crops. In the steppe areas far from markets it is grown extensively. In the more humid areas close to markets, such as northwest Europe, it is grown intensively. In the United States, wheat gives way to maize (corn) eastward as the climate becomes moister and the growing season lengthens. Corn is a fussier crop than wheat, and where climate allows, it is more profitable. In the eastern part of the corn belt, near large market centers, corn is shipped in bulk to the market since it does not have far to go. Farther west, more of the corn is fed to cattle or hogs, thereby converting it into a higher unit-value product which can better bear the cost of long shipment. This choice of alternatives is also affected by the relative prices offered in the market for corn and hogs or beef, which continually vary. The corn-belt farmer, especially if he is in a median location close to the break-even point between shipping corn and shipping hogs or cattle, has to keep in mind what is called the *corn-hog ratio*. At certain times market prices may make it more profitable for him to ship his produce as hogs, at other times as corn. Specialization of this kind, or of most kinds, is not possible in areas which lack cheap transport. Soil or land in such areas, despite its regional variations, is more uniformly used to grow cereal and other high-calorie food crops for immediate local consumption by dense subsistence populations. Market distances are therefore less relevant. In commercial economies land near the market will generally have a high rent because of competing uses and because market nearness minimizes transport costs and maximizes net profits. Intensive use of the land is therefore more necessary and is in keeping with the principle that a scarce or high-cost resource or production factor will be used intensively. Thus there is a pattern of increasingly intensive use of the land in commercial economies as the market is approached. Figure 20 attempts to show in generalized form the effect of market distance on agricultural land use, eliminating all factors except distance and using lines instead of the overlapping transition zones which are closer to reality.

Even in subsistence agricultural systems, shifting or sedentary,

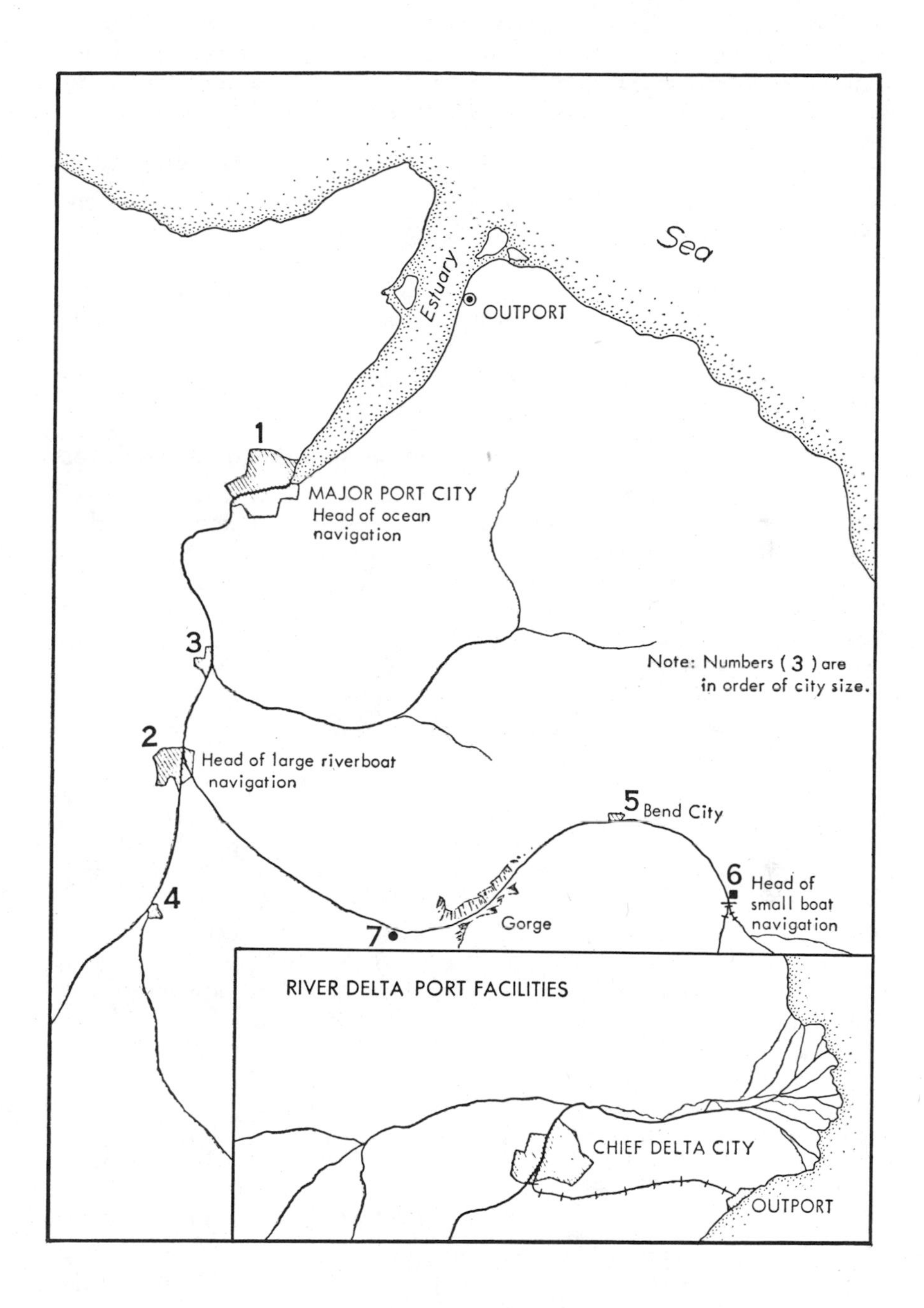

Figure 22. Hypothetical Settlements Along a River.

land nearest to a temporary or permanent village will be worked most intensively, and the village will tend to be located in the optimal soil and water area. Land farther out will not repay as well intensive cultivation. The von Thünen model may thus apply in such situations also.

The Cost of Transportation

Distance may be measured in miles, *real distance*, or in costs, *economic distance*. Economic distance is certainly the more accurate and important measure in analyzing the nature and distribution of most of man's activities or in determining the uses to which a given soil may be put. Cost more than miles gauges market accessibility. Transport costs depend on a variety of factors. The volume, constancy, and direction of the flow, the type and capacity of the carrier, and the nature of the route are fully as important as the real distance.

The Competition of Other Areas

A soil or an area may be well suited to a particular use, but another elsewhere may be better, or the agricultural system in another area may produce at lower cost due to more extensive methods, local advantages, or more efficient operation. Still another area may be nearer to the market, in miles or in economic distance. This competitive relationship is likely to be changeable, especially as transport costs change. Commercial wheat growing in the steppe, for example, was not possible until cheap rail and sea transport was developed. Extensive wheat-growing methods in the steppe, developed in the nineteenth century and dependent on mechanical cultivation, made possible a low unit-cost of production despite the marginal climate and the relatively low yields per acre. The agricultural development of the steppe and of the American Middle West were partly responsible for the relative decline of agriculture in old England and for the widespread abandonment of farms in New England beginning about 1830. Much of New England is now essentially a series of clearings in a secondary forest where once were cultivated fields whose old stone-wall borders punctuate almost any walk in New England woods. Much of the Connecticut valley, however, is still cultivated profitably, thanks to fertile alluvial soil and nearness to markets which combine to keep Connecticut-valley agriculture competitive. European agriculture has undergone similar changes as technological development and transport improvements made overseas agricultural production competitive on the European market.

The Available Technical Means

Some soils require special techniques and cannot be used, or their use will be restricted, until these techniques are available. Much of the steppe could not be cultivated profitably until stronger plows and tractors to break the tough sods and mechanical harvesters for extensive farming of large fields were developed. Mechanical cotton pickers helped to push cotton growing onto the soils of California at the expense of fruit crops. New crops, introduced from other areas or developed to tolerate certain conditions, may make previously neglected soils usable. Irrigation and drainage can revolutionize the agricultural potential of a soil. The availability of particular techniques varies widely. In preindustrial areas, technical limitations help to restrict the variety of possible soil uses.

The Pressure of Population

Population pressure has already been discussed as it may affect some farming practices. Where population pressure is high, where the economy includes limited non-agricultural segments, and where food is not cheaply available from elsewhere, land will be used intensively whether it is particularly fertile or not. Marginal land will be used which in the absence of population pressure would be neglected. The agricultural system will tend to concentrate on crops yielding a high calorie return per acre, such as rice or potatoes. The place of grazing animals for meat will be small because much more land is needed to support animals, in proportion to the human food they provide, than to raise an equivalent amount of food from grains or other field crops.

LAND NEAR THE MARKET

Where market considerations are dominant, special soil characteristics may become important. The middle- and south-Atlantic coast of the United States, for example, is generally not a particularly fertile soil area, but its sandy soils are mainly light in texture, well drained, easily tilled and fertilized, and they warm up early in the spring, especially since they lie near the sea. Their nearness to the large market centers of the east coast makes it profitable to use these soils for perishable truck crops, which find good drainage, fine texture, and early warming especially advantageous qualities in soil. Truck crops from the sandy coastal plain mature earlier in the year and command a premium price on the market when the heavier clay soils have not yet produced a crop. Denmark and parts of Holland, with soils similar to the Atlantic coastal plain in the United States and similarly located with respect to large market centers, have similar agricultural specialties. In the United States, with its highly developed system of fast low-cost transport, country-wide competition from fruit and vegetable crops grown in Florida and California limits the scope of market gardening around cities in other parts of the country. Local market gardens, however, have a net advantage in summer, when they can put many perishables on the local market at a lower cost and in a fresher condition. Market-garden areas are distinguished from truck gardens or truck farming because they are right next to the city, whereas truck farming may be carried on wherever transport gives it access to a market at bearable cost. Market gardens have to use what soil they find in the city environs, but even if the soil is poor, nearness to the market outweighs most soil disadvantages and makes it profitable to invest in soil improvements. Chance pockets of good soil around cities will often attract market gardening or horticulture and greenhouses. These specialties are agriculture of very high value, and the good soil pockets may thus escape the outward spread of housing considerably longer than surrounding suburban areas with poorer soils. When the price of land reaches a critically high point as the city expands, agriculture yields to housing, retailing, manufacturing, or other non-agricultural land use which will pay a higher price for land than farming can pay.

With soils, as with the environment as a whole, physical factors do not operate alone and may often be of relatively minor importance. The significance of location and of spatial relations especially is clear in accounting for land use and may often override the importance of the local soils. Technological or demographic (population) factors may also be of controlling importance. But the environmental influence on man is seldom limited to the immediate physical or economic opportunity which it offers or denies to him. The ramifications of a particular agricultural system and its requirements, or of the general economic level to which a soil may contribute, will often go far beyond a simple physical or local relationship to involve widespread aspects of the society as a whole. This is generally true of the over-all relationship between man and his environment.

Questions for Further Study and Discussion

1. Where are the local market gardens which supply the city or town where you live? Why? On what kinds of soil are they located? Are other factors involved in their siting? Have any local market-garden areas been eliminated by other land use in the past ten years? Can you observe any consistent change or gradation in the value or type of the crop as the city or town is approached?

2. If you keep a garden, what kind of soil does it have? What do you do to increase its productivity? What should you do? How does your problem as a gardener differ from that of a Dakota wheat farmer? a market gardener near a big city? a corn-belt farmer? an Indian peasant farmer?

3. What can the new chemical soil conditioners do for a soil? Why are they not more widely used, especially in "hardpan" areas? Are they likely to be more widely used in the future, and if so where and why?

4. Can one establish a significant correlation between productive soils and climate and tenancy or slavery in all cases? (Use examples other than those cited.) Between soil variations in your state or area and income levels, quality of housing, political sympathies, or other characteristics?

5. Why should land which can profitably grow "fussy" crops be expensive?

6. What deltaic cities other than New Orleans have site problems related to the local soil? Why should laterite occur in the area of Freetown?

7. What crops are grown in the valley of the Connecticut River? Why? How are these crops related to soil, climate, and markets?

8. In what detailed respects is wheat growing in the steppe areas of North America or Australia extensive, as compared with intensive agriculture in the central valley of California, for example?

9. What is the average yield of wheat per acre in England? In Australia? (Consult the figures of the Food and Agriculture Organization of the United Nations, or an agricultural atlas.) Why does England import wheat from Australia? Define *comparative advantage.*

10. What are the advantages of growing different crops on a given farm? What are the disadvantages of single-crop agriculture? What examples can you find of single-crop agriculture? Why is it so widely practiced? Is it for the same reasons in each case?

11. Why do "most tropical soils tend to be poor"?

Selected Samples for Further Reading

Albrecht, W. A. "Soil Fertility and Biotic Geography," *Geographical Review*, XLVII (1957), 86–105. An important and useful discussion of the relation between climate, soil fertility, and food value of crops produced, with some of the implications for man.

Bennett, H. H. *Elements of Soil Conservation*. 2nd ed. New York, 1955. One of the best general surveys of this broad problem by an outstanding authority on soils and land use.

Buckman, H. O., and Brady, N. C. *The Nature and Properties of Soils*. 6th ed. New York, 1960. An excellent though rather technical comprehensive treatment.

Bunting, B. T. *The Geography of Soil*. London, 1965. The process of soil formation, classification problems, and detailed studies of great soil groups.

Carter, G. F., and Pendleton, R. L. "The Humid Soil: Process and Time," *Geographical Review*, XLVI (1956), 488–507. A valuable discussion by two authorities of some of the basic aspects of soil formation, including the close relation between podzolization and laterization.

Eyre, S. R. *Vegetation and Soils*. See the reading list for Chapter 9.

Goguel, F. *Géographie des élections françaises de 1870 à 1951*. Paris, 1951. A more careful work than Gosnell's, but with a broad sweep.

Gosnell, H. F. *Grass Roots Politics*. Washington, 1942. A semi-popular survey of American voting habits which includes some discussion of the relation between soils and politics.

Howard, A. *The Soil and Health*. New York, 1947. A sample of the extensive literature on the relation between soil and nutrition and the need for organic fertilization–more objective than many such studies.

Keller, F. L. "Soil Erosion–Some Influences of the Social Environment," *Geographical Review*, XLVI (1956), 114–16. Examples from Latin America of the connection between erosion and differing forms of culture and associated land use.

Sinclair, R. "Von Thünen and Urban Sprawl," *Annals of the Association of American Geographers*, LVII (1967), 72–87. A brief survey of the von Thünen theory as applied to urban areas, and analysis of how recent American growth has altered the pattern.

Van Riper, J. E. *Man's Physical World*. 2nd ed. New York, 1971. Chapter 16 of this standard text includes a clear and detailed survey of soil classification schemes and their evolution over the past couple of decades, a process still going on.

United States Department of Agriculture. *A Manual on Conservation of Soil and Water*. Washington, 1954. A simple, brief handbook of good soil practices, including material on the importance of proper vegetation cover and use of water, with many excellent photographs.

———. *Soil*. 1957 Yearbook of Agriculture. Washington, 1957. Many of the articles in this volume supersede articles of similar title in the earlier soils volume of the Department (1938 Yearbook), but many are on different topics. The most useful include the following:

Dean, L. A. "Plant Nutrition and Soil Fertility," pp. 80–84.

Hendricks, S. B., and Alexander, L. T. "The Basis of Fertility," pp. 11–16.

Richards, L. A., and Richards, S. A. "Soil Moisture," pp. 49–50.

Russell, M. B. "Physical Properties," pp. 31–37.

Simonson, R. W. "What Soils Are," pp. 17–30.

———. *Soils and Men.* 1938 Yearbook of Agriculture. Washington, 1938. This volume contains a number of useful summary articles, clearly written by authorities. Although in a few respects some of the articles are becoming slightly out of date, several of them are still among the best brief surveys available. The most relevant articles include the following:

Byers, H. E., *et al.* "Formation of Soil," pp. 948–78.

Hambridge, G. "Soils and Men—A Summary," pp. 1–46.

Kellog, C. E. "Soil and Society," pp. 863–86.

Leighty, C. E. "Crop Rotation," pp. 406–30.

Watters, R. F. "The Nature of Shifting Cultivation," *Pacific Viewpoint,* I (1960), 59–99. A valuable detailed survey of recent research, with a good discussion of the interplay of environment and culture and an extensive bibliography.

11
Water and Man

ASIDE FROM THE OCEANS, which are not generally considered as water resources because they are not yet economically drinkable or usable for most industrial, agricultural, or domestic purposes, water is present in the ground, as vapor in the atmosphere, and in surface bodies such as rivers or lakes. Atmospheric water may be just as important as ground or surface water; the difference between dry air and moist air can be critical to man and to his livelihood. Atmospheric moisture also has an important bearing on temperature as well as on rainfall and on evaporation rates, which may be as significant as the actual amount of rainfall. Water in the ground is contained or circulating in the soil as moisture, the amount of moisture depending on the climate and on the texture and structure of the soil. Below a certain level under the surface, the ground is usually saturated. This level or line is called the *water table* (see Figure 21), and it follows roughly the contour of the surface, varying from place to place as the soil or rock varies and as the supply of water varies. Ground water from this saturated zone is the source of wells and springs. Where layers of impervious rock retain water, pressure may accumulate on an underground slope as a result of gravity, and a well drilled, or a spring naturally occurring, at a low point on this slope may flow or spout without pumping, to form an artesian well or spring as illustrated in Figure 21. The saturated zone also helps to maintain moisture in the soil above it, through capillary action, just as coffee rises in a sugar lump when only one end is wet. If the water table is normally low, or if for any reason it substantially falls, as it regularly does to some degree in the drier season of the year, natural vegetation and human activities on the surface may be seriously affected. Man himself, through his wells and pumps, may often be the main factor in lowering the water table.

WATER AND SETTLEMENTS

With the growth in the last hundred or hundred and fifty years of many cities of a million or more, and with the coming of the modern

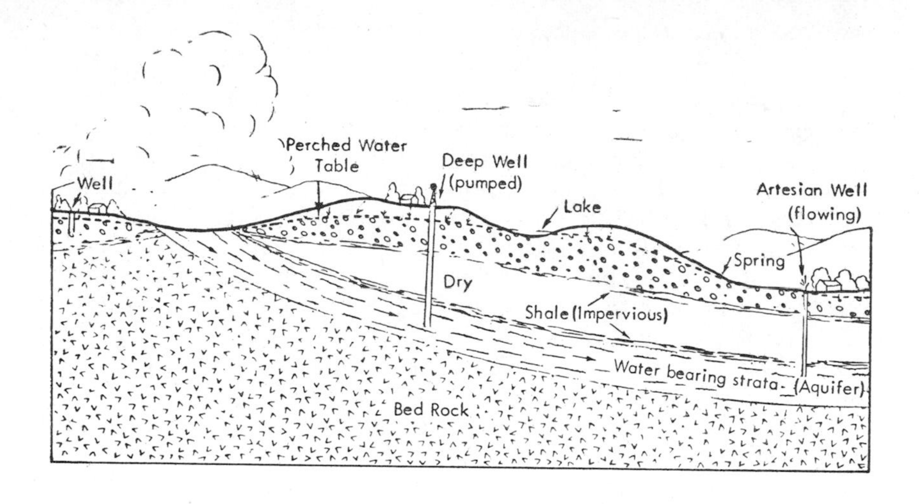

Figure 21. Diagram of Ground Water Occurrence.

industrial revolution in large factories consuming enormous quantities of water, ground water has become a less important source than rivers or lakes. Even though it may have to be brought from a great distance, river or lake water provides the more constant and higher volume of supply which large-scale urbanization and industrialization require and which wells alone could not meet. But before the urban and industrial revolution, and in many of the less densely settled or less industrialized parts of the world today, ground water supplies most of man's water needs unless he happens to live close to a lake or river. The importance of the water table for agriculture has, of course, remained in any case, but in earlier periods or in the less industrialized parts of the world now, human settlement and all of man's activities may be severely limited by what the local ground water can supply. The original sites of many towns and cities were fixed, in whole or in part, by local water. The drier the climate, the more limiting this factor was and is; water is not universally available. In a desert, of course, this limitation is painfully obvious, and there settlement is confined to the few scattered places where the water table will support wells or springs, or where a river rising in a humid area flows through the desert.

A map of world population provides clear evidence of the limitations imposed by water. Look at the area in central Asia known as the Tarim Basin in Hsinkiang, politically part of China and lying north of India and Tibet. It is a nearly rainless desert and is largely uninhabited. But around its rim are several spots of very dense population, even though beyond these spots there is still almost uninhabited country for hundreds of miles. The Tarim Basin is a desert partly because it is so far from the sea and partly because it is virtually surrounded by some of the highest mountain ranges in the world, which effectively shut out or milk dry any moisture-bearing winds which penetrate this far. There is, however, some rain or snow on the mountain flanks and crests, and this feeds a number of streams which tumble down onto the desert floor of Hsinkiang. They quickly lose themselves in the desert, but their water can be tapped at the desert margins by damming or diverting the streams as they drop down out of the mountains, or by drilling wells into their weakening underground courses farther out onto the plain. This is the basis for the spots of dense population.

In this case, water, like most environmental influences, provides a permissive rather than a causative basis. It allows human settlement and agriculture in the small oases and denies them in most of the rest of Hsinkiang. Soils in the oases are productive enough under irrigation and strong sun to yield a surplus of agricultural goods for trade. Because water is available, settlement in the oases is also able to perform another profitable function as way stations for the land trade routes between China and India and between China and the West. The most important route westward from ancient China ran through Hsinkiang; it was called the "silk road" from the most profitable commodity which it carried, and it is still used to some extent for more mundane goods. The route had to pass through the oases in order to obtain water and food, and the oases supported much of their populations by servicing and using the trade route. On the map of world population are several other places where dense populations stand out in an otherwise empty desert. One can assume that the principal reason for these sharp distinctions is water. In Egypt it is the Nile, rising in wet equatorial Africa and flowing across the desert to reach the sea, supporting in its desert valley one of the densest concentrations of human settlement in the world. In the Libyan Desert it is local oases drawing on ground water, and in the Wasatch oases of Utah in the Salt Lake City area, mountain streams dropping down onto the desert basin, a situation similar to Hsinkiang's.

MODERN URBAN-INDUSTRIAL WATER USE

It is not necessary to look at the desert for evidence of water's influence on settlement, although water's commanding role is especially clear there. In moister climates, too, water is usually an important locating factor for settlement and economic activity, either as ground water from wells, or in the form of rivers or lakes. Preindustrial man needed much less water but could not move it around easily or cheaply and had to locate his settlements and tailor his economic activities in accordance with local ground or surface water. Industrial man needs huge amounts of water; he has the means to move it considerable distances, but he does not thereby avoid the economic aspect of the problem. His agricultural, industrial, and urban establishments are largely fixed, or at least they can be moved only with great difficulty or expense. The nomad can move when local water is exhausted, but a city, an irrigated farm, or a steel mill cannot; they must pick their sites or adjust to them more selectively.

As individuals, Americans probably use fifty times as much water per capita for their daily needs as did their great-grandfathers. They wash themselves and their property, including cars, more assiduously, mainly because large piped water supplies and household plumbing make it much easier and cheaper. Average Western urban water use in 1970 was about 150 gallons per capita per day for domestic purposes; the average city of half a million produced in 1970 over 1800 tons of solid waste and about fifty million gallons of sewage. Daily total water consumption for all uses in the United States in 1972 was well over sixteen hundred gallons per capita, a staggering fact. But in industrial areas by far the biggest water consumer is manufacturing, and for the United States as a whole, the largest single consumer of water is irrigation. Personal or individual water consumption, although it has increased rapidly, is proportionately small. The figure for industrial use is difficult to make precise, since most industrial operations reuse water more or less continuously, whereas irrigation uses water directly only once. New York City uses nearly a billion and a half gallons of water a day. Personal consumption can be somewhat curtailed if necessary in an emergency, but it is much more difficult to reduce industrial consumption unless the industry itself is shut down. Most manufacturing processes require water, and some use enormous amounts. As much as 65,000 gallons, or about 260 tons, of cold, fresh, and relatively soft water is used or reused for cooling and flushing in the making of one ton of steel. Sea water is too corrosive, and some fresh water is too hard or leaves deposits which obstruct pipes. Most of the cooling and flushing water is commonly reused, but in the case of one steel plant near Youngstown, Ohio, which reused water from a small river on which it was located, it was found that the temperature of the river as a whole had been raised to 140° Fahrenheit.

When the growth of population on the west coast of the United States and the strategic and transport problems of the Second World War combined to bring about the construction of a steel plant somewhere west of the Mississippi, Utah was chosen because it was the nearest area to the west coast as a whole which had adequate cheap deposits of coal with iron ore nearby. But most of Utah is a desert, and the most convenient location for the plant, Salt Lake City, could not use the salty water of the lake. The site finally picked, Provo, was a compromise among the factors of coal, iron ore, market and labor, transportation, and water, of which water was the most controlling. Provo lies near the shore of a large lake, fortunately fresh. But even in humid areas such sites are relatively limited and must be chosen carefully. This is one of the reasons for the location of iron and steel manufacturing along the Great Lakes and on the Ohio and Delaware rivers.

Plants generating electricity from steam-driven turbines, usually the largest source of electric power for most big cities, are even more closely tied to water. In total they account for over half of the industrial water use of the United States. Power plants are almost always located directly on a river or lake, from which they take water to cool their condensers. The chemical and paper-making industries are also especially large water users. But few industrial operations can afford to locate their plants where ample cheap water is not available. Water is just as essential a raw material to them as anything else they use, and it may be more limiting, since manufacturing needs such large amounts of water that the costs of transporting the water are very high. Industrial demands for water are still increasing rapidly and in the United States are expected to be 200 per cent greater in 1980 than in 1950.

Water is also widely used as a cheap and effective means of dispersing industrial and household wastes. But a polluted stream is less valuable as a source of water for domestic or industrial use, and pollution may also kill commercially valuable fish and may ruin the stream as a recreation area. Upstream users in this case can effectively destroy the water for downstream users and may even destroy it for competing uses of their own. Alternate means of waste disposal, treatment of wastes before dumping them into streams, or subsequent treatment of the water before reuse are all expensive, but they are clearly essential as the pressure of competing water uses mounts. Many American rivers have been ruined by pollution. Water is far from being unlimited or free for any of the uses to which it is put. It must be husbanded and allocated like any other finite resource whose use involves the danger of exhaustion and requires the consideration of competing demands.

Most modern large cities have outgrown their original water supplies and have had to reach out long distances for water. Los Angeles is often cited as an example of this problem, sending covered aqueducts over two hundred and fifty miles to the Owens and Colorado Rivers; in the early 1970's water also began to flow seven hundred miles to the city from the Feather River in northern California. This costs money, and it is done only because the Los Angeles area has advantages for settlement, agriculture, or manufacturing which other parts of the southwest lack. As a huge city on the edge of a desert, Los Angeles has a special problem, but it is really no worse than New York's in the humid east. New York must compete for its billion and a quarter gallons a day with many of the other large eastern cities, since its search for water has largely exhausted what is readily usable in its own tributary area, and New York has begun to tap supplies which Boston and Philadelphia respectively regard as theirs. New York, among many other cities the world over, experiences recurrent water crises and may be saved from them eventually only by the development of a process to distill or freshen sea water at a bearable cost. This point has now very nearly been reached, but it seems likely that whatever happens, city water may soon have to be more rigidly metered and sold like any other valuable commodity and at a higher price than is now paid for it.

Fortunately most large population clusters and most manufacturing have grown up in the well-watered parts of the world, and the total economic advantages of these places are also great enough to pay, if necessary at high rates, for the solution of most water problems which arise, although many headaches and maladjustments may have to be overcome and many disputes settled between competing users. In the semi-arid areas of the world, disputes over water rights are chronic; those between competing national states for the use of water from the Jordan River, which flows through Lebanon, Syria, Israel, and Jordan, are an example. On any river upstream use may be detrimental to downstream users. These problems arise, for example, with the Colorado River in the United States, between Arizona, California, and Mexico, and there have been continual disputes over water between India and Pakistan.

IRRIGATION

Agriculturally, ground and surface water are important for maintaining soil moisture naturally, or for augmenting it artificially through irrigation. Water for irrigation may be derived from below the water table through wells, or it may be supplied from a greater distance from surface bodies if the agricultural system can pay for it. Where wells drain too much out of the ground and the water table falls below the limit of the wells, agriculture dependent on wells bumps up against a clear physical limit unless it can supplement its water supply from elsewhere. Lowering the water table may also damage the soil. In parts of semi-arid lowland California where there has been a great expansion of irrigated agriculture based on wells, the water table dropped so far that it was below sea level and allowed an invasion of sea water which poisoned the soils. This was a somewhat unusual situation, but lowering the water table may have equally serious effects on vegetation and crops in many semi-arid areas, where there is the greatest need for wells.

Irrigation from any source pays only if it can support an adequately productive agriculture. Of the total area in western United States which could physically be irrigated by wells or aqueducts only a part actually is irrigated. Only in these places has irrigation so far been profitable, because their soils, topography, and location are favorable enough to make it economically possible for them to pay for the water. They can, and must, grow high-value crops or must get a high return from the land to pay for the irrigation; fruit, fancy vegetables, cotton, sugar beets, or multiple crops of alfalfa as cattle feed all produce a large cash return per acre. Cotton and sugar beets are specially soil-demanding; irrigation for them must be confined to soils which will profitably support them.

Irrigation is also characteristic of many subsistence agricultural systems, as opposed to the commercial system of Anglo-America or most of Europe. In a subsistence economy where there is relatively little sale of the product and most of it is consumed by the producer, irrigation "pays" in terms of the pressure of population on the land and the need to maximize production per acre. Irrigation is confined to areas where preindustrial techniques can get the water, and in these areas population is very much denser than elsewhere, because irrigated farming can support it. Irrigation may also be applied to areas which are not usually dry but which suffer to a degree, like all of the world, from seasonal and cyclical fluctuations in climate. In well-watered eastern United States, for example, despite its relatively reliable rainfall, irrigation can almost always increase agricultural yields by avoiding the damage caused by dry spells.

WATER POWER

There are two associated aspects of water resources which should be considered here: power and transportation. Hydroelectric power is sometimes spoken of as a "free" resource; we shall consider in some detail in Chapter 13 why this is not so. The world contains a great deal of falling water which might be made to produce power, but only a small fraction of it is so used. The reasons for this are in part physical or technical, but far more they are economic. Potential water power is not evenly distributed over the earth, since it is of course related to climate and to relief; some areas largely lack it, others are well supplied. But where it is physically plentiful it may not be economically usable. It is estimated that about 20 per cent of the total world water-power potential lies in one area of the Congo, but almost none of this 20 per cent is used. On the other hand, Japan, with a little over 1 per cent of the total world potential, uses almost all of its 1 per cent and has thousands of times the developed water power of the Congo.

The Example of Grand Coulee Dam

An American example may help to explain the difference between Japan and the Congo and differences within the world as a whole in the use of water power. There are hundreds of physically possible power sites even in the relatively dry country west of the Mississippi, but only a few of these have been developed. The largest development is Grand Coulee Dam. It is a huge project and cost a great deal of money to build. Why was Grand Coulee chosen in preference to other sites? The source of power is the Columbia River, which derives its water from the relatively humid parts of the Canadian and American Rockies and then flows across semi-arid or desert country, where the dam lies, to the Pacific Ocean west of Portland. Although the Columbia is fed primarily by snow-melt water and is characterized by peak flow in spring and early summer and low volume in winter, its flow is more dependable from season to season and from year to year than is the case with most snow-fed or desert-crossing rivers. The Columbia also carries relatively little silt, since so much of its water is derived from snow or from reasonably well-forested areas; silt can damage turbines and can fill up storage reservoirs behind dams. Storage reservoirs are designed to minimize the effects of fluctuating stream flow, but the problem is easiest and the costs lowest when stream flow fluctuates least. If fluctuation is too great or possibilities for water storage too limited, it may not pay to build a power installation which can operate at capacity only for a few months when stream flow is adequate. At Grand Coulee the river is somewhat confined in a wide gorge, and since it is also losing altitude rapidly it flows with considerable force. Behind the dam site the gorge widens further and easily allows for the backed-up waters which now form a huge reservoir, so that the flow of water through the turbines at the dam can be kept constant. Physically, this is a good place for a power dam.

But power must be sold, and few people live in the arid country around the dam. Power cannot be transmitted economically in most cases for more than about three hundred miles. Beyond that distance too much power is lost from the lines, or too many booster stations are necessary, to make transmission pay. Power also cannot be stored economically in any significant amounts. Since there are high fixed charges on the dam in the form of interest on the huge investment and of maintenance, the plant must produce and sell power continuously. It cannot afford to shut down or to run at a fraction of capacity. Hydroelectric establishments, in other words, are economically possible only if they have within a range of roughly three hundred miles a large market with a high and constant demand for power, or what is called a high constant load factor.

Well within a three-hundred-mile radius of Grand Coulee lie nearly all of the large cities and manufacturing establishments of the far west, north of California. The urban and industrial area of Seattle-Tacoma is about 160 miles away, Portland about 250, greater Vancouver (as large as Seattle) about 210, and Spokane about 75. Because Grand Coulee can deliver large amounts of power to them, many new, heavy power-using industries have been attracted to these cities and thus have helped to cut the costs of power down. The larger and more constant the sales of power, the lower the unit costs (the cost of one unit of power). Population growth in the cities served by Grand Coulee, based largely on new or expanded industry, has thus had a tendency to snowball. Industries which require large amounts of power (for example, aluminum refining and the smelting of certain ores) have become prominent in these cities since Grand Coulee power has been available and are now an areal specialty. As pointed out in Chapter 2, one person employed in a basic industry is usually accompanied by about two people employed in the service industries which bake the bread, clean the streets, staff the professions, operate department stores, or supply materials to the basic industry. Counting family dependents, it is estimated that one worker in a basic industry usually results in about eight other people dependent on him, serving him, or dependent on those who serve him and his dependents. In this way Grand Coulee has contributed substantially to the growth of population in the Pacific Northwest. This may be attributed in part to water resources, but weight must also be given to the force of location or access and to the existing or potential markets which the dam could reach. Little manufacturing has gone to the dam itself. Manufacturing has settled in established urban centers which were there before and which are based on a variety of other advantages, including labor force, transportation, access to productive hinterlands, and favorable site.

Most of Grand Coulee's revenue comes from power sales, but this was not enough to meet the costs of the dam, especially at the time it was planned and built in the 1930's, when future power sales, the impact of the Second World War, and the growth of population were greatly underestimated. Some of the water is also used for irrigation, and sales of irrigation water were expected to help meet the cost of the dam. Irrigation of this part of the Columbia plateau can be justified in part because the soils are exceptionally good. The parent rock is volcanic lava, and the semi-arid climate has allowed a fertile and well-structured soil to develop. Although water from the Columbia has to be pumped up out of the gorge about three hundred feet to the level of the surrounding plateau, irrigated agriculture on the plateau is very productive. The irrigated farms concentrate on high-value crops, mostly multiple crops of alfalfa and vegetables, and they realize a high return per acre. Thus soil, as well as location and site, helps to explain why a dam at Grand Coulee was possible.

Most modern dams are multipurpose projects whose costs are met through a variety of uses. Power and irrigation are usually the most important, but flood control, the maintenance or improvement of navigation, water for household or industrial use, and recreation on the artificial lake which may result behind the dam are also common benefits. Usually it takes a combination of uses to pay for, or to justify, the building of a big dam. Some of them, like recreation or flood control, are hard to put a money value on, but they must be considered. The general effect on the economy of the area of all these things taken together is also worth something; it will increase the population, raise the average income level, and make the area a pleasanter or more profitable place to live. These indirect benefits should be calculated and some of the dam's costs charged against them. The extent to which all or any of these products of a dam can profitably be used (and thus whether the dam can be built) will depend on where the dam is and what kind of country and people lie around it or have reasonable access to it. Water resources in this sense clearly involve much more than the physical fact of falling water.

WATER TRANSPORT

Water transport is usually the cheapest way to move bulk goods. Friction is slight, and costs of building or maintaining the right-of-way are low compared to most forms of land transport. But most important is the fact that the carriers, or their capacities, can be very much larger than on land without being proportionately expensive to operate. Generally, the larger the carrier, the cheaper its unit haulage costs, assuming the carrier can consistently make up a full load. There is a point of diminishing returns beyond which increasing the size of the carrier does not bring lower unit costs, but the *economies of scale* (the bigger, the cheaper) work in favor of water transport. If carriers, by land or water, can be built to facilitate mechanical loading and unloading or to haul only one type of cargo which can make maximum use of the carrier's capacity, costs are further reduced. The largest single cost item of transport, especially for large shipments, is often loading and unloading, known as *terminal costs*, rather than the actual movement along the line, known as *line costs*. Water carriers often achieve economies in this respect, particularly in specialized ships like tankers, or ore, coal, or grain boats. General cargo is more expensive to load and unload and to carry, but even for general cargo most water carriers have terminal-cost and line-cost advantages over most land carriers. Another important factor is length of haul. Since terminal costs may often be greater than line costs, long hauls are often disproportionately cheap, if a full load is available for the entire distance. Long hauls usually cost less per mile than short hauls, especially for large carriers. Most ocean hauls are necessarily long, and ton-mile rates (the rate for carrying one ton one mile) can therefore be very low.

Waterways and Population

Population and economic activity along waterways is also of course due in part to transport and access advantages. Economic opportunity is enhanced where goods can be assembled or distributed cheaply. This fundamental association is clearly revealed in a world map of population. Although some unnavigable, desert, or arctic rivers seem to be avoided by population, a larger-scale or more detailed map would show that most of these rivers have attracted many more people than the surrounding areas. Many densely settled areas do not coincide with rivers or lakes (in several cases these areas are close to the sea instead), but most surface water is associated with dense population, and vice versa. Notice on the world population map how the Amazon and its tributaries stand out from the equatorial rainforest through which they flow, and how the Congo and the Niger do much the same, although there the population clusters are in scattered blobs. The Parana in Argentina is notable on this map, as are the Mississippi, Ohio, Tennessee, Delaware, Hudson, and Mohawk in the United States, together with the St. Lawrence and the Great Lakes. In Europe the Rhine and the Elbe are easily followed, and the plain of the Po in northern Italy stands out. The Nile has already been discussed, but notice its parallels in the Tigris-Euphrates in Iraq, the Amu Darya and Syr Darya in Russian central Asia, and the Indus in Pakistan; they are parallels because they are all, like the Nile and the Columbia, *exotic* rivers, rising in a wet place and flowing across a desert to reach the sea. Exotic rivers are seldom very important for transportation, being largely unnavigable except for small boats as a result of seasonal fluctuations in volume and of silting, but they bring water and alluvial soil, a potent combination.

In the north of Asia notice how population follows the great Siberian rivers. This taiga area, like the rainforest, is characterized by low mobility, and even though the rivers are frozen most of the year and may flood for much of the rest, they were the best means of circulation available until the building of the trans-Siberian railway. The railway is now followed by even denser population, but the broader zone of somewhat sparser settlement at greater distances from the rail line, especially in the eastern half, is associated with the Amur and with the rivers of Siberia. In China, India, and the Indo-Chinese peninsula which lies between them, there are almost no relatively densely settled areas divorced from rivers except in narrow coastal strips, and nearly every major river valley has attracted a lion's share of the population. This is not only for transport reasons; alluvial soil (particularly valuable in tropical lateritic soil areas), level land, and water for irrigation are also provided by the rivers. But mobility in these mainly preindustrial parts of the world tends to be limited, and cheap easy circulation by river is an important factor in attracting population. In Japan and most of oceanic southeast Asia mountains are too general and watersheds too small to allow navigable rivers to any extent. But, except for interior Borneo and New Guinea, the sea is close and is a vital factor in the Japanese, Javanese, and Philippine economies.

Waterways and Economic Development

Before the industrial revolution and the building of railways, water transport was the only economically practical carrier for bulk or low unit-value goods moving beyond a few miles. Transport by cart or pack animal may within fifty miles double the cost of the goods carried unless they are of very high value per unit. In preindustrial periods or areas, settlement and economic development are particularly tied to waterways. Rivers were the almost universal means of exploration and settlement in colonial areas: witness the roles, to give only a few examples, of the Hudson-Mohawk, Ohio, Mississippi, Connecticut, James, Columbia, and St. Lawrence in North American history, the Amazon, Congo, Nile and Volga for their areas, and in the East the Ganges, the Irrawaddy, and the Yangtze, on which Western economic colonialism so massively rested. With the coming of the modern industrial revolution there was a tremendous increase in the volume of exchange and a great need for cheap carriers to move huge amounts of low unit-value goods like coal, iron ore, or foodstuffs. Water transport has increased greatly in absolute terms where industrialization has spread, even though with the building of railways water transport has declined relatively. But the Rhine, for example, still carries a greater volume of traffic than the combined railways which parallel the river, although German railways as a whole haul more goods than internal waterways.

Some smaller areas which originally were nourished economically by waterways, Cape Cod, tidewater Virginia, or parts of the Ohio valley, for example, have declined when railways or roads short-circuited traffic around them. There are ghost towns and rotting wharves in many such areas, where a few miles away the railroad and the automobile have stimulated new towns. Most cities founded before 1860 necessarily grew up around or along the water routes which served them. These older parts of many cities have now become relatively less prominent and may even be decaying as transport and economic activity have increasingly centered on railroad termini or motor routes. But frequently the railway termini, at least for freight, have grown up in close association with the original waterfront, where cheap access and break-in-bulk points are concentrated. Passengers can now travel more quickly than by water, and their termini are often separated from the old waterfront where the entire life of the city originally centered. Such local or small-scale readjustments should not obscure the continuing importance of water transport, however. In the three greatest centers of manufacturing, Japan, Europe (including the Ukraine), and eastern North America, naturally available water routes have been a strong advantage, an important reason for early industrial leadership, and a continuing stimulus to economic growth.

Settlement on Rivers

Settlement and economic activity are seldom uniformly spread along a river. The effects of the fall line have been discussed in Chapter 3, and similar cases can be found in many areas outside the eastern United States. There are usually a variety of other break-in-bulk points along a river where carriers change. At or near the mouth is perhaps the most common, since most rivers are not navigable for ocean-going ships much above their mouths. The most economic ocean carriers tend to be the largest, and it usually does not pay to haul goods all the way across the ocean in a boat small enough to use even a big river. Most rivers are settled particularly densely near their mouths, where level land and fertile alluvium are also likely to be concentrated. An atlas map of the United States or of the world will show how close a relationship exists between large cities and river mouths. In Europe the mouth of the Rhine supports three large cities (Rotterdam, Amsterdam, and Antwerp), as well as most of the Dutch economy, though not entirely on the basis of Rhine traffic. But there is usually a succession of heads of navigation farther upstream: one for ocean-going vessels, one for large river boats, one for smaller river boats, and so on. At each well-marked break, which may be a falls, or rapids, or a narrowing of the river or its valley, a town or city may arise, taking advantage of the opportunities which any break-in-bulk point provides. Such a city has to consider access to the land also, and it is therefore unlikely to be in a gorge, for example, but downstream from the gorge where there is a wider productive hinterland and easy access routes to it. On the Yangtze River, Hankow, nearly three hundred miles downstream from the Yangtze gorges, is a much larger city than Ichang, which lies at the downstream entrance to the gorges. Hankow is surrounded by fertile lowland and is also at the mouth of the Han River, one of the Yangtze's major tributaries. Similarly, Montreal is larger than Cornwall, Ontario, where the gorge and rapids of the St. Lawrence begin, and Montreal is also straddled by the mouths of the Ottawa River where it joins the St. Lawrence.

Confluences are favorite settlement sites, as Hankow and Montreal suggest. The main stream is likely to be less navigable above the confluence than below, and there may also be a change of carriers between the larger stream and the smaller one. There will certainly be

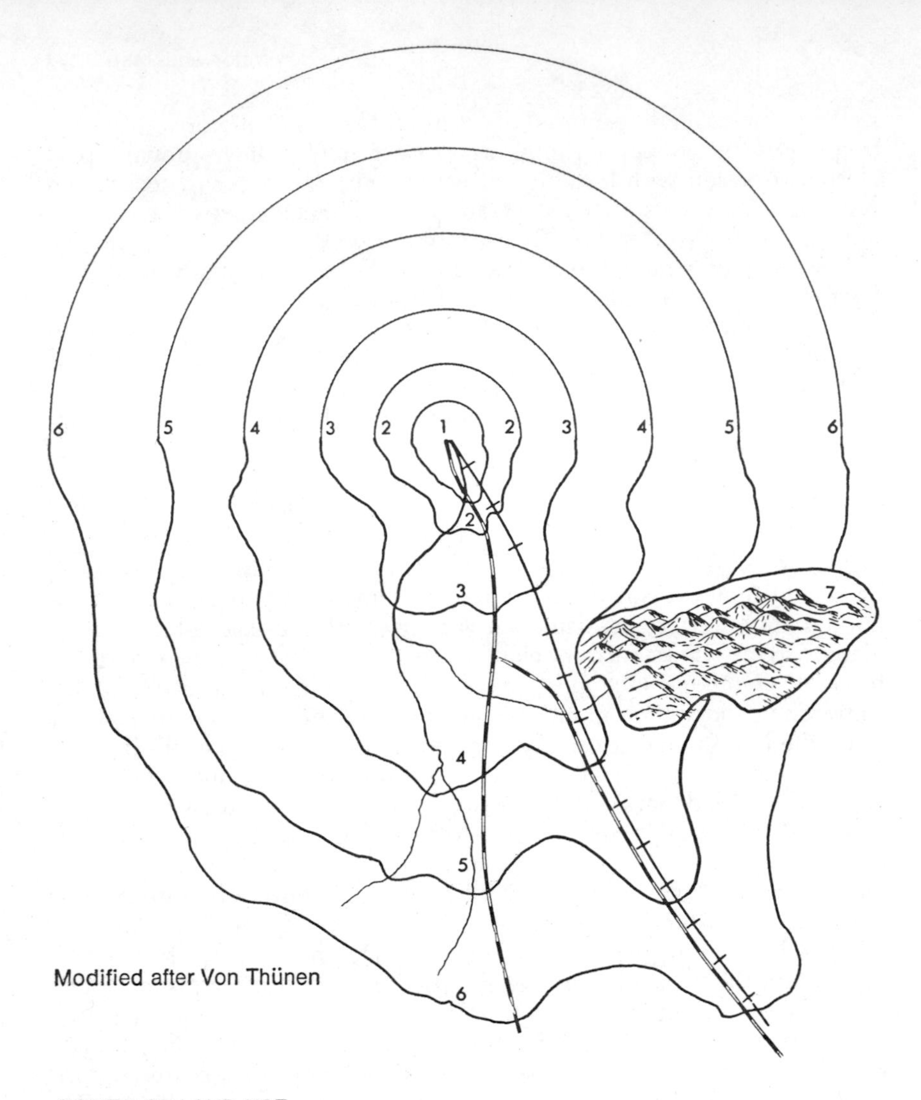

ZONES OF LAND USE

1. Non-agricultural urban area
2. Poultry, horticulture and market gardening
3. Market gardens, fruit and fluid milk
4. Dairying, truck farming and other types of intensive agriculture
5. General farming, grain, hay and livestock
6. Open range or forest-range in areas of semi-arid plains, forests in moist hilly areas
7. Minor grazing and dairying, forestry on steep slopes

(Lower half of figure suggests modifications induced by variations in landscape.)

Figure 20. Hypothetical Land Use Zones Around a City as a Function of Distance.

a focus of routes and opportunities for access in different directions, and the rivers may have helped to create a broad productive lowland. Philadelphia, where the Schuylkill flows into the Delaware, is an example; Pittsburgh, where the Allegheny and Monongahela rivers join to form the Ohio, is another. Cincinnati, St. Louis, Omaha, and Minneapolis-St. Paul are also confluence cities. As with many cities at river mouths, however, the exact point of the confluence may be avoided; often it may be subject to floods, and higher ground may be available near enough to take advantage of the confluence.

At the river's mouth depth may be adequate for ocean-going ships for a space upstream, and it is cheaper for ships to go up as far as they can as long as they do not run into expensive delays. Silting and sand bars may be less of a problem upstream or on a downstream tributary of the main stream, as at Shanghai or Calcutta, than at the edge of the delta. The lower delta may also be more subject to floods or may have permanently sodden or salty soil. New Orleans is about a hundred miles from the mouth of the Mississippi, although there is no question of a fall line near New Orleans, nor near Hamburg, which is over fifty miles from the mouth of the Elbe.

Many cities so placed have had problems, because the size of ocean-going ships has increased particularly rapidly since the 1890's. Some cities have established *outports* at the new downstream head of ocean navigation where the larger ships load and unload, especially passenger ships, which are not only larger than freighters in most cases but also carry a cargo which is in a hurry. Cuxhaven is the outpost for Hamburg, Bremerhaven for Bremen (haven is a variation of *hafen*, the German word for harbor or port), Le Havre for Paris (Rouen), and Gravesend-Tilbury for London.

London illustrates another point on a river where settlement tends to cluster: the site of a bridge or ford. Often the lowest point on a river which can be bridged will also be the head of ocean navigation, as it is with London, with several of the cities of the fall line in eastern United States, and with many other urban centers elsewhere. But any bridge site is likely to be a good break-in-bulk point and assembly-distribution center, since land routes will be bunched together there in order to get across the river. "Bridge" or "ford" in the name of a city may reveal one of the city's site advantages. It does, for example, with Hartford, Connecticut, Oxford and Cambridge in England, Klagenfurt and Innsbruck in Austria, Saarbrücken in the Saar, and Bridgeport, Alabama. Familiarity with maps and the study of place names can be fascinating.

Questions for Further Study and Discussion

1. How can one explain in detail the distribution of irrigated lands in the United States? What does each major irrigated area grow? Are they generally crops bringing a high monetary return per acre? How can one account for the areal differences in crops?

2. Where is the power plant or plants in the city or town where you live? Why? Is water the only locating factor?

3. How can one explain the large development of hydroelectric power in Switzerland? Japan? Sweden? France? Italy? Is load factor the only determinant? How can one account for the proportionately much smaller development in Canada, Brazil, Spain, China, and Australia? Are the reasons the same in each case?

4. How can one explain the interrupted concentrations of population along the Congo as contrasted with the more even settlement pattern along the Amazon?

5. What examples can you find, in addition to those mentioned, of confluence cities? bridge-site cities? cities at various upstream heads of navigation? outports?

6. Where rivers have winding courses, cities often find advantages in particular locations along the bends. What location or locations? Why? (Consult an atlas for examples.)

7. Why should there be particularly complex disputes over the use of water from the Colorado River? Why should India and Pakistan have had conflicts over water use?

8. Why are steam plants using coal and steam-driven turbines to produce what is called *thermal* power usually the largest source of electricity for most cities? Why is hydro power not the largest? Is any thermal power used in cities close to large sources of cheap hydro power? Why or why not?

9. Why is population so dense in the desert valley of the Nile? Is water the only factor?

10. Are most *exotic* rivers navigable? Why, or why not? What precisely does "navigable" mean?

11. How many cities are there in the world with populations of more than half a million which are not on navigable waterways or on the sea?

12. Some river mouths have not attracted a large city, or the largest city may be located off to one side of the river or upstream. Find several instances, apart from those cited, and account for each. Can you establish any generalizations? Why in detail is New Orleans so far from the mouth of the Mississippi?

13. The Nile is a big and important river. Why is Cairo not an important port? Why is there no important port even at the river's mouth?

14. Why is most manufacturing in the United States concentrated east of the twenty-inch isohyet?

15. One often hears statements such as "Country X has a potential hydroelectric production of fifteen million kilowatts." What uncertainties or variables are involved in or concealed by such a statement?

16. What is a "power grid"? How does it affect the possibilities of transmitting electric power? What other possibilities may there be of transmitting power economically over distances greater than three hundred miles? How would such developments affect the possibilities of tapping presently unused hydro power?

Selected Samples for Further Reading

Ackerman, E. A. *Water Resources in the United States*. Resources for the Future, Series No. 6. Washington, 1958. An excellent brief summary.

Alexandersson, G., and Norstom, G. *World Shipping*. New York, 1964. Subtitled "An Economic Geography of Ports and Seaborne Trade," this comprehensive volume provides detailed data and analysis of all important ports, cargoes, carriers, and ocean routes.

Chorley, R., ed. *Water, Man, and Life*. London, 1969. A multi-author general survey, especially good on hydrologic systems and the systems approach.

Eckstein, O. *Water Resource Development: The Economics of Project Development*. Cambridge, Mass., 1958. Benefit-cost analysis, indirect benefits, and an excellent summary discussion of the economic aspects of irrigation, hydro power, navigational improvements, and the allocation of joint costs.

———, and Krutilla, J. *Multiple Purpose River Development*. Resources for the Future. Baltimore, 1958. A somewhat broader work than the preceding item, but mainly directed at cost analysis.

Fox, I. *National Water Resource Policy Issues*. Resources for the Future, Series No. 4. Washington, 1957. A discussion of government and local policy in the United States in recent years.

Hodge, C., and Duisberg, P. C., eds. *Aridity and Man: The Challenge of Arid Lands in the United States*. Washington, 1963. American experience with arid areas, a series of case studies, and problems for the future.

Kuenen, P. H. *Realms of Water*. London, 1955. Trans. from the Dutch. A physical survey of water on the earth and in the atmosphere, including an excellent chapter on glaciers.

Olson, R. E. *A Geography of Water*. Dubuque, Iowa, 1970. The physical nature of water, and problems of its management.

Thomas, H. E. "Changes in Quantities and Qualities of Ground and Surface Water." In W. L. Thomas, ed., *Man's Role in Changing the Face of the Earth*, pp. 542–63. A good brief summary of the hydrologic cycle and of man's effects upon ground water, including an extensive and valuable bibliography.

United Nations. *Proceedings of the United Nations Scientific Conference on the Conservation and Utilization of Resources: Vol. IV, Water Resources*. New York, 1951. Papers and discussions dealing with irrigation, water power, flood prevention, navigation, ground water, and urban-industrial water consumption, by a variety of international authorities.

United States Department of Agriculture. *Soil*. 1958 Yearbook of Agriculture. Washington, 1958. This large volume contains two useful articles on irrigation:

Criddle, W. D., and Haise, H. R. "Irrigation in Arid Regions," pp. 359–67.

Quakenbush, T. H., and Thorne, M. D. "Irrigation in the East," pp. 368–77.

———. *Water.* 1955 Yearbook of Agriculture. Washington, 1955. Like the other Yearbooks of Agriculture, this is a large volume of collected articles by various authorities on a number of topics related to the title. This volume includes articles on water needs, water sources, water and soil, water and forests, irrigation, drainage, wildlife, and problems of the future.

Walton, K. *The Arid Zones.* Chicago, 1970. Adaptation of plants, animals, and men to the water-poor environments which cover over one-third of the earth's land surface.

Weigend, G. "River Ports and Outports: Matadi and Banana," *Geographical Review*, XLIV (1954), 430–32. A good brief statement of the growth of outports and the role of river ports, with special examination of two ports on the lower Congo.

———. "Some Elements in the Study of Port Geography," *Geographical Review*, XLVIII (1958), 185–200. A study principally of port hinterlands.

White, G., ed. *The Future of Arid Lands.* Washington, 1956. Collected papers from an international "Arid Lands" meeting, including studies of a variety of areas in different parts of the world and extensive bibliography.

———. "Industrial Water Use," *Geographical Review*, L (1960), 412–30. A critical review of recent literature, from both North America and Europe, stressing that the problem of water availability is more one of an effective supply system than of actual reserves.

12
Landforms and Man

THIS LAST OF the five interrelated elements of the environment is perhaps the only one which can rival climate as an influence on man. Landforms may arbitrarily but conveniently be divided into four categories: plains, hill, plateaus, and mountains. Hills are not only lower than mountains but are usually more rounded and more gently sloping; plateaus are simply elevated plains. Elevation, however, is less important than *local relief*. Local relief includes the degree of slope of the ground, how much difference there is in feet between the highest and the lowest parts of an area, and how much of the area as a whole is level or sloping. A highly elevated plateau may be more usable than a low-lying area with heavy local relief. Many areas labelled mountains may actually contain extensive sections of low relief, and many plains may be interrupted by broken country.

MAPPING RELIEF

Most relief maps of large areas rely on altitude tints to show elevation above sea level, often using green for elevations under a thousand feet and darkening shades of brown, purple, or red for the higher elevations. Altitude tints may obscure great variations in local relief. They may also, as in the case of the North American Great Plains, suggest topographic barriers where none in fact exist. The map maker must draw his lines between altitude tints at some more or less arbitrary elevations. If he must map a large area with widely varying elevations, each color category must cover hundreds or even thousands of feet within which there may be considerable local relief which the altitude tints cannot show. A map of India, for example, must include both the Himalayas and the Ganges valley, but at normal page size such a map would be very difficult to read if it used more than eight or ten categories of altitude tints; each color would thus have to represent one to three thousand feet, and each would therefore conceal local variations which might be only a few feet less.

More accurate relief maps of smaller areas are made using multiple *contour lines*, as in Map 1 in Chapter 2 on page 29. (*A contour line* connects points of equal elevation.) The smaller the *contour interval*, or the difference in elevation between two contour lines, the more accurate the map, and the more difficult it is to read. The smaller the variations in local relief, the smaller the contour interval which can be used and still leave the map reasonably clear. Where small local variations in relief are especially important, such as in irrigated areas or the centers of cities, a contour interval of five feet or less may be used. For larger areas or areas with heavy local relief, an interval of fifty or a hundred feet or more is more appropriate. What is called a *vertical profile* may also be constructed on the basis of a contour map, as illustrated on Map 1. For a large area small-scale contour maps are almost illegible or unworkable unless generalized so much that altitude tints might just as well be used instead. The best small-scale relief maps of large areas make use of a combination of altitude tints and what is called *plastic shading*, now used in most atlas maps of landforms. This visually suggests the landforms without losing any accuracy and makes the map much more vivid and less flat looking. But no small-scale relief map can be accurate in detail, and it can give only a highly generalized picture.

RELIEF AND POPULATION

Nevertheless, a comparison of a world physical map with a population map will demonstrate a clear association between people and plains. Only in the tropics or in South Africa do large numbers of people live in the highlands. This is not so much because in the tropics climate is pleasanter in the highlands, but because the highland climate offers greater economic opportunity. Desert, arctic, and equatorial plains are avoided; climate overrules landforms. Otherwise there is remarkable consistency. Japan, which the world map makes seem an exception, actually is not; the Japanese highlands are sparsely settled and the lowlands almost uniquely crowded, although most world maps are too small to show this adequately. The combination of altitude tints and small scale plays tricks in a few other places, most notably in southeastern Spain and in the Chinese province of Szechuan; both actually contain extensive areas of low relief. Notice on the population map how the Alps, the Pyrenees, the Rockies, the Himalayas and their many associated ranges, the Caucasus, and many lesser mountains stand out as empty areas. Why is there this general avoidance of mountains and concentration on plains?

Lowlands and Mobility

Lowlands have a number of obvious but vital advantages. Circulation and mobility, conditions of economic development and of resource use to support large populations, are maximized on plains. Mobility means that areal specialization can proceed—that each place can maximize its particular advantages and at the same time receive the special products resulting from the advantages of other places. If everyone had to grow his own grain, bake his own bread, and make his own alarm clocks, we would not get very far. Many mountain people do have to do roughly this, but on the plains cheap circulation helps to make it unnecessary. The division of labor (job specialization), dependent upon exchange, is a universal basis of economic growth. Manufacturing, especially heavy manufacturing, is absolutely dependent on cheap exchange. Its bulky raw materials cannot be moved around otherwise, nor can its finished products be distributed. Where manufacturing is established on the plains, dense population is likely to result, as basic labor and to provide services. This further increases the market and attracts commercial production of many kinds, including commercial agriculture in the market area. Mobility over the plains is not due only to the absence of topographic barriers. Rivers flowing across relatively level areas are more likely to be navigable than mountain or hill streams; the significance of cheap water transport has already been discussed. In many cases the rivers have helped to create the plains by eroding away the original mountains or by depositing silt to fill up ancient valleys or coastal seas. Actually, many of the great plains areas of the world either fringe the sea, where rivers have extended the land by deposition, or lie open to the sea. Coastal plains have attracted the largest blocks of dense population, as shown on the population map, where the circulation advantages of the sea are available as well as the mobility of the lowlands. A great deal of the world's heavy manufacturing is concentrated on coastal plains.

Plains Agriculture

Manufacturing (including the services attached to it), and exchange or commerce are two of the three major economic activities requiring and contributing to dense population. The other is agriculture. The remaining activities—herding, hunting, forestry, fishing, and mining—are either extensive in nature, are concentrated in marginal or severe climates, or may have no close connection with major settled areas. Lowlands offer overwhelming advantages for agriculture. The many problems of cultivation on slopes are avoided. The soils of the plains are nearly always superior and may also have benefited from alluvial deposition. Outside the tropics and if the sea is near, as in many plains it is, climate is likely to be moderate, in both rainfall and temperature, compared to that in the highlands. In the two most heavily peopled lowlands in the world, in India and China, these agricultural advantages are the primary attraction and basis for settlement. But even in industrial Europe and North America the agricultural virtues of the plains are of great importance. Commercial agriculture depends on market access just as manufacturing does, and access and markets are maximized on the plains, where most people live. A dense population requires a productive agricultural system, whether its nature is commercial or subsistence. The availability of surplus food and the presence of the relatively dense populations associated with productive farming areas are further attractions for most manufacturing, which must obtain and feed its labor and must have access to consumers as well as to raw materials. As the eighteenth-century historian Edward

Gibbon put it, in a perceptive appreciation of the geographic concept of spatial relations, "The first and most natural root of a great city is the labor and populousness of the adjacent country, which supplies the materials and subsistence of manufactures and of foreign trade."[1] Populousness and productivity from agriculture or manufacturing, plus easy mobility, mean that lowland areas are also likely to be generating centers for trade and foci of trade routes.

MOUNTAIN SETTLEMENT

One may ask why anybody bothers to live in the highlands. Most people like their native places, and often the more different from adjacent places the more passionate the attachment. If one's native place is economically poor, its other virtues are certain to be extolled, or poverty itself may be made a virtue. Mountains are physically distinctive places especially sharply differentiated from lowlands, and those who live in the mountains often assume that they have the better of the bargain, even if not for economic reasons. In practice, mountains make good refuge areas, and many mountain people originally came there because they were pushed off the better lands by stronger or larger groups. As with the rainforest, security is the result not only of physical barriers but of the physical and economic poverty of the mountains, which attract few contestants. Religious, social, and political dissidents may also swell the mountain population, and there are always a few hardy souls who honestly prefer the highlands as a place to live. In many mountain areas human settlement is too old, or too little is known of its origins, to say for certain whence came its inhabitants or why they went there. But with few exceptions outside the tropics, they are not there because the mountains offered them greater economic opportunity than the plains.

Poverty may also be reflected in emigration. Most mountain areas send out a constant stream of people seeking better opportunities elsewhere. Brigandage or organized raiding of more favored places may also be a highland specialty, in Europe in the past, in some other parts of the world still. Mountain conditions—clan rivalries or the need for defense against pressure from the lowlands—may encourage the development of military prowess, and economic hardship may provide an incentive for the use of this prowess. Mountain and plains peoples seldom get along well together, and this further tempts the highlanders. The hills are frequently sources of trouble, either for the lowland trading town or farmer who may be raided or for the government anxious to preserve order. The shoe may be on the other foot when the economic and political power of the plains reaches out to subdue or control the mountains. History is full of such two-way struggles, and where highlands and lowlands meet, the landscape may be marked with fortresses and battle sites, such as Stirling in Scotland and Carlisle in north England or Carcassonne in southern France. The spate of

---

[1]Edward Gibbon, *The History of the Decline and Fall of the Roman Empire* (London, 1787), Chapter 71.

movies, short stories, and novels about military adventures in the North-West Frontier Province of India or on the borders of Scotland are good melodramatic reminders of such struggles. The plains people are far from guiltless in this and have often regarded the highlands as fair game.

Local Mountain Advantages

In some cases highlands may have net advantages over adjacent lowlands. This may be the result of a critical degree of security provided by the landforms, or, as in the tropics, of the more permissive climate associated in some places with elevated areas. Both factors are involved, for example, in the Plateau of Iran, which has a much denser population than most of the surrounding desert lowlands; the lowlands have suffered not only from greater aridity but from repeated conquest and devastation at the hands of nomadic raiders. The Plateau of Iran offered a degree of protection from such depredations, and the climatic map shows much of the plateau as distinct; rainfall is only a little greater than on the lowlands, but temperatures and evaporation rates are lower. Such situations are uncommon outside the tropics, however, and are in any case mainly limited to plateaus, where relatively level land is available to support people. In Tibet, which is only part of a vast mountain region, there are few advantages, although the population map shows a small cluster of people in the southeastern part of the country. Actually, out of the unimpressive total of fewer than one-and-a-half million people in Tibet (which is no more than one would expect), well over 80 per cent live in the valley of the Brahmaputra, or Tsang Po as it is called in Tibet, which coincides with the small cluster on the world population map. In the Asian tropics, highland locations do not seem to have attracted settlement as much as they have in other tropical areas, and most of the lowlands, far from being avoided, are extremely densely populated.

MOUNTAIN ECONOMIES

Mountain cultures and economies are as distinctive as their physical base. Heavy manufacturing and field agriculture, except in chance pockets of level or fertile land, are either excluded or severely limited. This may leave scope for dairying or for the raising of sheep and goats for wool or skins. Sheep and goats will tolerate slopes and relatively poor pasturage better than cattle will. Their meat may be consumed locally but can seldom be exported; it is too low in value per unit of weight and too perishable to be carried profitably by the transport available. The wool is often made into cloth before export for the same reason, and transport costs also explain the making of cheese. Handmade manufactures are prominent, as in Switzerland, due to a long slack season and to limited local resources in both a physical and economic sense. Value added by manufacture is stressed, and hand-woven woolens, for example, may command a high price, although they seldom return a very high hourly wage or profit to the maker. Bulky or perishable exports are difficult. Grains tolerant of the poor soils or harsh climates common in mountain areas may be grown on more level patches of ground for local subsistence or for animal fodder. To be exported, grains must usually be converted into a higher-value form as animals, for example, or as whiskey; Scotland and Kentucky still illustrate this point. In very high mountains specially adapted animals may be used: llamas or alpacas in the Andes, yaks in the Himalayas and Pamirs. All of these animals are physiologically suited to high altitudes, and indeed to not thrive below about ten thousand feet. In Tibet and the high Andes they are the preponderant source of food, clothing, transport, fuel (in the form of dung), and exports. One of Tibet's exports for many years, in addition to wool, was yak tails, high in value by weight and virtually imperishable, to be used as fly whisks in China and India and as false beards and Santa Claus whiskers in the United States and Europe. This is a good measure of mountainous Tibet's poverty and of its basically subsistence economy, in common with many mountain areas.

Tree Crops

Slopes need something which will hold the soil, and bush or tree crops may be as good as grass or may thrive as well. Two particular tree or bush crops grown almost exclusively in the hills are tea and coffee. Both tolerate slopes, like good drainage, and profit from the coolness and moisture of higher elevations. With tea this is more a matter of economics: the lowlands are taken up by the more profitable and more particular crops. Where it is grown on hill plantations, as in India or Ceylon, tea may support a good many people locally. In China and parts of Japan it is a largely untended, casual crop, visited by lowland people only for occasional pickings. Coffee is grown almost entirely in the tropics, since the tree and its fruits are damaged or killed by frost. The coffee tree dislikes excessive heat, however, and so prefers the highlands, where slopes do not especially bother it. Tea and coffee are probably the most important commercial crops of the mountains over the world as a whole. If the sparsely settled mountains are forested and the adjacent densely settled lowlands denuded, as is often the case, lumbering may be another source of livelihood, but the transport problem is particularly pressing. Areas where the economic distance to market is great must concentrate for export on high unit-value goods. Timber is not high in value by weight, and the ruggedness and remoteness of mountain areas often mean that timber there is not exportable, unless it can be made into something locally which can bear the heavy transport costs. Wooden Swiss cuckoo clocks can, and so can Panama hats, which despite their name are made (the best of them) in mountainous, remote Ecuador, where there is need for an export whose high unit-value comes mainly from human labor, since there are few other local resources.

Water Power and Minerals

Potential water power and mineral deposits are commonly found in mountains, but only under certain circumstances do they contribute much to the density of population or to the level of the local economy.

In many cases they may not be usable at all. Water for uses other than power may often be derived from mountains for the supply of a less humid nearby lowland, as in the Mediterranean Basin as a whole, southern California, the Wasatch piedmont oases of Utah, or Hsinkiang. But the water is used in the lowlands, not in the mountains, important as the mountains are in this respect for the support of the plains people. The conditions under which water power becomes a usable resource have already been considered in the preceding chapter, and it should be clear that while falling water may be localized in the mountains, this means that it will often be beyond economic reach of an adequate market. In Japan it is not; plains and mountains lie close together, and over-all distances are relatively small. A huge nearby industrial market provides a high constant load factor. In Switzerland hydro power can be developed to support the diversified Swiss industries. In France, Germany, Italy, and Scotland, mountain sources of power are near enough to large markets. But in many mountain areas this is not the case; even where it is, the power is likely to be shipped out to where the people are rather than being used in the mountains, and this is in fact the almost universal pattern. Switzerland in this as in other respects is an exception to the rule.

It is much the same with minerals, which are often associated with or exposed by the folding and faulting of the earth's crust, the basis of most mountains. Especially if they are minerals low in value per unit of weight, like coal, they may be too remote, measured in economic distance, to be usable. If they can be brought out at a bearable cost, they will be brought *out*, not turned into manufactured goods in the mountains where they occur. The lowlands have a virtual monopoly of the other advantages for manufacturing, and they draw minerals to them.

Tourism

Mountain areas are often attractive as resorts for people from the crowded plains, especially where the plains are hot in summer. But situation is usually more important in accounting for the presence of tourists than the grandeur of the scenery or the cool weather. The Poconos and the Catskills, hardly in the mountain class, attract annually millions of vacationers from nearby New York and Philadelphia, while the infinitely more impressive ranges in Ladakh or Alaska rarely see visitors. Nearly every even slightly elevated or hilly area near a densely populated urbanized lowland derives a large and increasing share of its income from tourism, especially where transport facilities are well developed and average lowland incomes are high enough to permit luxuries such as vacations.

MOUNTAINS AND TRANSPORT ROUTES

The cost of transportation is a crucial factor in mountain economies. Transport routes generally avoid mountains, not so much because of physical barriers but because the mountains are relatively unproductive. Sparse settlement on a low economic level cannot support a railroad and may not even be able to support a road. It is true that land transportation, especially the bulk carriers like railroads, must be sensitive to grades. But where there is an economic basis for exchange, land routes and even railroads can afford to penetrate mountains despite the physical barriers and consequently increased costs, as for example in the coal areas of the Appalachians or through Switzerland. For the most part, however, transport routes penetrate mountains primarily to get to the other side. When roads or railways cross mountains, they try to combine a physically easy route with a commercially productive one. This is the basis of mountain passes. Some passes have been chosen principally or solely for physical reasons, but in most cases economic considerations are also involved, if not in the existence of the pass, then in the degree to which it is used or in the choice of alternate routes, depending on what productive or less productive lowland areas are connected by the routes. Settlement is not likely to concentrate in the pass itself; there will be few functions for a settlement to perform aside from sweeping the snow or landslides off the road or providing St. Bernard dogs for benighted travelers. A better place for settlement is at the lowland approaches to the pass, where there may be a need for a break-in-bulk point, for a strategic or military post, or simply for an assembly and distribution center where lowland routes are gathered into a knot before ascending the pass. Good examples of such settlements would include Milan, Denver, Cheyenne, Peshawar at the approaches to the Khyber Pass in Pakistan, and Toulouse on the French side of the Pyrenees. Trade is generated and handled on the lowlands.

Generally, transport routes penetrate mountains in order to get across them, because the mountains separate complementary areas between which trade flows. Occasionally roads or railways will do more than get across the mountains and may spread their lines out to haul minerals down to the lowlands. There was a time, for example, before about 1930, when there was a greater development of railroads in the mineralized areas of the Rocky Mountains of the United States than on the surrounding arid lowlands, which by comparison offered fewer cargoes. But this simply emphasizes the point: transport routes go where there is something to haul, whether it is in the mountains or not. Most mountains have little to haul, and this contributes to the high cost of hauling that little, whether it has to go on a man's or an animal's back, or on a one-train-a-week narrow-gauge railroad.

Transport Flow and Transport Costs

A high and constant volume of flow helps to keep unit transport costs low, just as greater capacity of the carrier makes for lower unit costs. The larger the capital investment in right-of-way, carriers, and capacity, the greater the need to use these facilities to capacity in order to pay off the high fixed costs. A large flow is necessary to justify the initial cost of building a road or railway and to maintain it in profitable operation. It may cost a million dollars or more a mile to build a first-class railroad or a modern turnpike. Where a large and constant volume of traffic makes it possible to provide such facilities, ton-mile costs for railroad haulage, for example, may be only 1 or 2 per cent of the costs of human or animal porterage in areas where there is not enough present or potential traffic to support a railroad. Human porters and pack animals, which are common in many mountain areas, are the most expensive of all carriers because their capacities are so small. Mechanized carriers of large volume, such as railroads, benefit from the economies of scale. Where the flow is mainly a single bulk commodity, it may be possible, as with ships, to use cars designed especially for that commodity, which can be loaded and unloaded mechanically, and ton-mile costs will fall even lower. A train may haul 5,000 tons of general freight at $.02 per ton-mile or 8,000 tons of coal at $.01 per ton-mile, whereas a pack animal may carry 200 pounds at $.80 per ton-mile and a human porter 100 pounds at $1.50 per ton-mile. The economic range of porters is also severely limited. If the volume of flow is small or irregular on a road or railroad, the facilities and total capacity will have to be limited if fixed costs of maintenance are not to swallow up all the profits; the average unit cost of haulage will consequently be higher than where large and constant volume makes larger facilities and capacities possible. Where the volume is too low to justify even road building, exchange will have to depend on still smaller carriers, cheaper to build or to operate per mile but much more expensive in ton-mile costs.

Transport costs may also be affected by the predominant direction of flow. Where most of the weight and volume is moving in a certain direction (the "heavy" direction), it may be possible to offer lower rates as well as more frequent service in larger amounts in the heavy direction. A carrier may often offer lower rates for freight moving in the "light" direction, however, in order to attract business and to meet fixed costs, especially if the carrier must run in the light direction anyhow in order to pick up freight which it will then move in the heavy direction. It is expensive to operate a carrier empty, and *return-haul* arrangements and special rates often attempt to compensate in such situations. Mountains usually send out a greater weight and volume of raw materials than they import as manufactured goods, and may therefore pose return-haul problems for carriers, sometimes reflected in higher rates. Similar relationships elsewhere, between agricultural or raw-material-producing areas and manufacturing areas, as between western and eastern United States as a whole, may affect transport rates. Finally, mountains pose obvious physical problems for any carrier, which are also reflected in transport costs. For all the above reasons, mountain transport tends to be expensive and thus further limits the economic opportunities of the mountains.

MOUNTAIN CULTURES

Mountain cultures can perhaps be summed up in the word "backwaters," but this would be unjust to many areas, or at least to many aspects of them. Switzerland is obviously the reverse of a backwater, despite the persistence of the Romansch language. But Switzerland is virtually alone among highlands in the degree of circulation and outside stimuli which it enjoys. Generally, these are at a minimum in the mountains; while that may be eminently desirable for some tastes, it undeniably tends to preserve archaisms. In speech, dress, customs, attitudes, and every other cultural attribute, most mountain people are usually out of gear temporally with the lowlands.[2] This may often be a blessing, especially from the mountain point of view. But although isolation may foster local feeling or keep worthwhile traditions alive, it also tends to mean that change is slow in the mountains and that they lack the dynamic quality of the lowlands, where circulation and external stimuli are concentrated.

In the mountains of Kentucky and Tennessee people still sing Elizabethan folk songs long since forgotten in England, and their speech is said to be much closer to the speech of Shakespeare's England than is the contemporary Londoner's. The mountain fastness of the Pyrenees long preserved the Basques, whose origins have been obscured by time, but whose language and culture are unrelated to any in Europe. They once occupied a larger area, including adjacent lowlands, and the mountains probably became for them a refuge area, where they now support themselves mainly by sheep raising. The tiny state of Andorra, also in the Pyrenees, owes its continued independence largely to the mountains. In Tibet Buddhism (and a variety of archaisms) was preserved, although Buddhism has all but vanished in its Indian home and in China, where it once flourished. Mountainous Ethiopia, early Christianized by missionaries from Egypt, kept its religion, though surrounded for hundreds of miles by other creeds, and today it may be closer to some of the forms of fourth- or fifth-century Christianity than is Rome itself, where much has happened since. Ethiopia's political independence throughout the age of colonialism in Africa (except for the brief Italian interlude) and despite several attempted conquests is also attributable in large degree to the physical protection and undesirability of the mountains.

DISTANT EFFECTS OF MOUNTAINS

Aside from human use of or adjustment to highlands, mountains may also exert their effects upon other areas, sometimes at a considerable distance. In Chapter 8 some of the climatic effects of mountains in creating rain shadows on one side and heavy rainfall on the other have already been discussed. Mountains usually have a strong influence on temperature also, shutting out maritime or continental influences. The Alps, for instance, plus the adjacent Massif Central in France, help to account for the warm sunny winter climate of the Riviera; the Himala-

[2]A French study provides some interesting evidence, finding that in adjacent mountain and valley voting districts (*arrondissements*) in the French Alps, the valleys tended with remarkable consistency to vote left and the mountains equally consistently to vote right. The mountain districts demonstrated consistent attitudes in other respects also, characterized as conservative and traditionalist. (Simone Hugonnier, "Temperaments politiques et géographiques electorales de deux grandes vallées intra-Alpines des Alpes du nord: Maurienne et Tarentaise," *Revue de geographie alpine,* XLII (1954), 45-80.)

yas and Hindu Kush shield India from the cold outblowing winter winds of Inner Asia and help to give India a virtually unbroken growing season; the bitter winters of lowland Montana and Idaho or of eastern Washington may be blamed in part on the Cascades and the Rockies, which shut out the moderating influence of the sea.

Any lowland area surrounded by mountains, or with one of its approaches blocked by them, is likely to be profoundly affected as a result, quite apart from the climatic effects. Even so large a lowland area as India shows these influences in every part. The external mountain barriers are formidable and have contributed much to India's marked cultural isolation from the Far East. But there is one gap region, and its location has been of crucial importance for Indian civilization. Through the Hindu Kush, Sulaiman, and other ranges on the west are several usable passes, including the famous Khyber, plus a relatively easy lowland route along a narrow corridor between the mountains and the Arabian Sea. India's one easy external connection by land was thus with the West, and through these gaps poured a long succession of influences and invasions which have tended to fragment Indian civilization and which have given it in many respects a Western character. Within the Indian peninsula another set of mountain ranges, the Satpura and the Vindhyas, still marks a cultural line between the Aryan or Western-dominated north and the Dravidian or more indigenous south. In the refuge area of the Vindhyas themselves, primitive aboriginal tribes still live.

Much of the distinctiveness (and backwardness?) of Spanish civilization is due to the barrier imposed by the Pyrenees, which are particularly effective because there are few easy routes across or around them, unlike the Alps. Spain is mostly mountains, at best a plateau, and the Pyrenees merely reinforce general cultural and economic isolation, hence the old expression "Africa begins at the Pyrenees." Within Spain there is still no common language on the scale of most national units, and there are very strong local regionalisms; to be a Catalonian or a Castilian in speech and in origin is often more important than to be Spaniard. Political separatism, archaic culture, and limited economic development in the Balkans may also be attributed in part to mountain barriers. China, densely populated and productive as it is, owes much of the character of its civilization to the isolation which an almost unbroken wall of mountains has helped to create, even though they lie far from the center of the Chinese lowland.

Topography (relief) may also help to deflect or to channel movements of ideas and of goods. Settlement and transport follow the lowlands, substantially penetrating the highlands in most cases only to get to the other side and making use of the lowest or most convenient routes. Mountain barriers may thus focus trade and affect a very large area. New York City, as already analyzed, owes much of its size and importance to the Mohawk corridor through the Appalachians, which focuses trade between the Middle West and the Atlantic coast. Shanghai profits from having its river-basin hinterland surrounded on at least two sides by mountains, which help to funnel Yangtze trade through the city. On the Pacific coast of the United States, Seattle, Portland,

San Francisco, and Los Angeles are all close to easy passes or river valleys leading east across the mountains. These sites are the only ones west of the Rockies which support large cities.

LOCAL RELIEF AND URBAN SITES

On a smaller scale than has so far been discussed, slope and elevation are usually of local importance as one aspect of site. A city or town, for example, may choose an elevated site to avoid floods, to moderate climate, or to gain military protection. Defensibility was a major concern for nearly all settlements until relatively recently, since only a few favored areas were free from chronic civil disorder or invasion. Only within the last two or three centuries have settlements been able largely to ignore topographic protection, or to omit walls or moats if the city had to be sited on level ground. Although elevated sites continue to have some advantages in flood protection and for residential use, trade and manufacturing suffer on such sites, and most large modern cities avoid high or rugged ground, at least for their business centers, if level ground is available. Many cities on elevated sites have retained their old centers, often built around a cathedral or castle on the fortified hill, and have developed a new commercial, manufacturing, and transport center on the adjacent lowland. Where enough level land is not available, or where it is less favorably located (Seattle, San Francisco, Boston, Pittsburgh), land may be reclaimed from the sea or river to support manufacturing and transport, especially for large space users like railroads, which often arrived on the scene after most of the existing level land near the center of the city was already occupied.

Residential housing can be more tolerant of slopes and may even seek them, particularly high-rent apartments and high-value single-family houses. In most American cities the higher or more rugged the land in the suburban area, the greater its price per square foot and the more expensive its housing. Such land is away from the commercial and industrial nuisances of the city's center, which depend on access and therefore seek level ground. Higher ground, if it presents problems for bulk transport, may also have escaped deforestation and may still be wooded; it is certain to have a pleasanter climate than the lowland center of the city, especially in summer. If slope makes the area less accessible, so much the better for those who like privacy and are willing to pay for it.

Agriculturally, slopes may attract fruit or nut trees or grapes, which like good drainage, do not mind slopes, and may be able to maximize sunshine on hillsides facing south. The famous Rhine wines come from grapes grown on such hillsides in the cool climate of the Rhine valley, where the slopes facing north are usually uncultivated. Similar use is made of hillsides in most mountain areas. But such local advantages or uses of slopes should not draw attention from the larger fact that in general, highlands and poverty go together. Man is predominantly a creature of the plains.

Questions for Further Study and Discussion

1. How can one account for the fact that almost every American city, including those with relatively level sites such as Chicago, Cleveland, or Philadelphia, has found it necessary to make artificial land by filling in rivers, lakes, or the sea? What is this filled land generally used for? Why? Consult Map 7. Find similar data for San Francisco, Pittsburgh, Chicago, Philadelphia, Baltimore, Cleveland, Los Angeles, and New York.

2. Is there any general pattern of city location which would help to explain the development of cities near the lowland approaches to mountain passes? (See Chapter 3.)

3. Compare an atlas map of an area which you know well with what you know about the relief of the area. Compare this map with a larger-scale contour map of part of the area which you know best. Topographic sheets using contour lines are easily available for nearly every part of the United States at a scale of 1 to 62,500 or slightly smaller. If you cannot obtain them locally for your area, write to the United States Coast and Geodetic Survey, Washington, D.C.

4. Construct a vertical profile from a topographic sheet, using the contours and the vertical and horizontal scales; the vertical scale is provided by the contour interval. Construct a contour map of a hypothetical area of varied relief.

5. What examples are there in American history of conflict between mountain and plains people? In European history?

6. Why are there fewer distinctively mountain cultures or economies preserved in the United States than in Europe or Asia?

7. Are there any places in the world where manufacturing is localized in the mountains? Why?

8. What distant effects of mountains can you discern in the United States?

9. How many cities in Europe arose without any topographic protection in the form of an elevated site or of surrounding hills? How can one account for these cases?

10. Why is the economic range of porters limited? How does the economic range of motor trucks compare with that of railroad trains? Why?

11. What was the origin of the "Whiskey Rebellion" of 1794 in the United States? How were the grievances of the rebels settled?

MAP 7

MADE LAND AND LAND USE IN DOWNTOWN SEATTLE

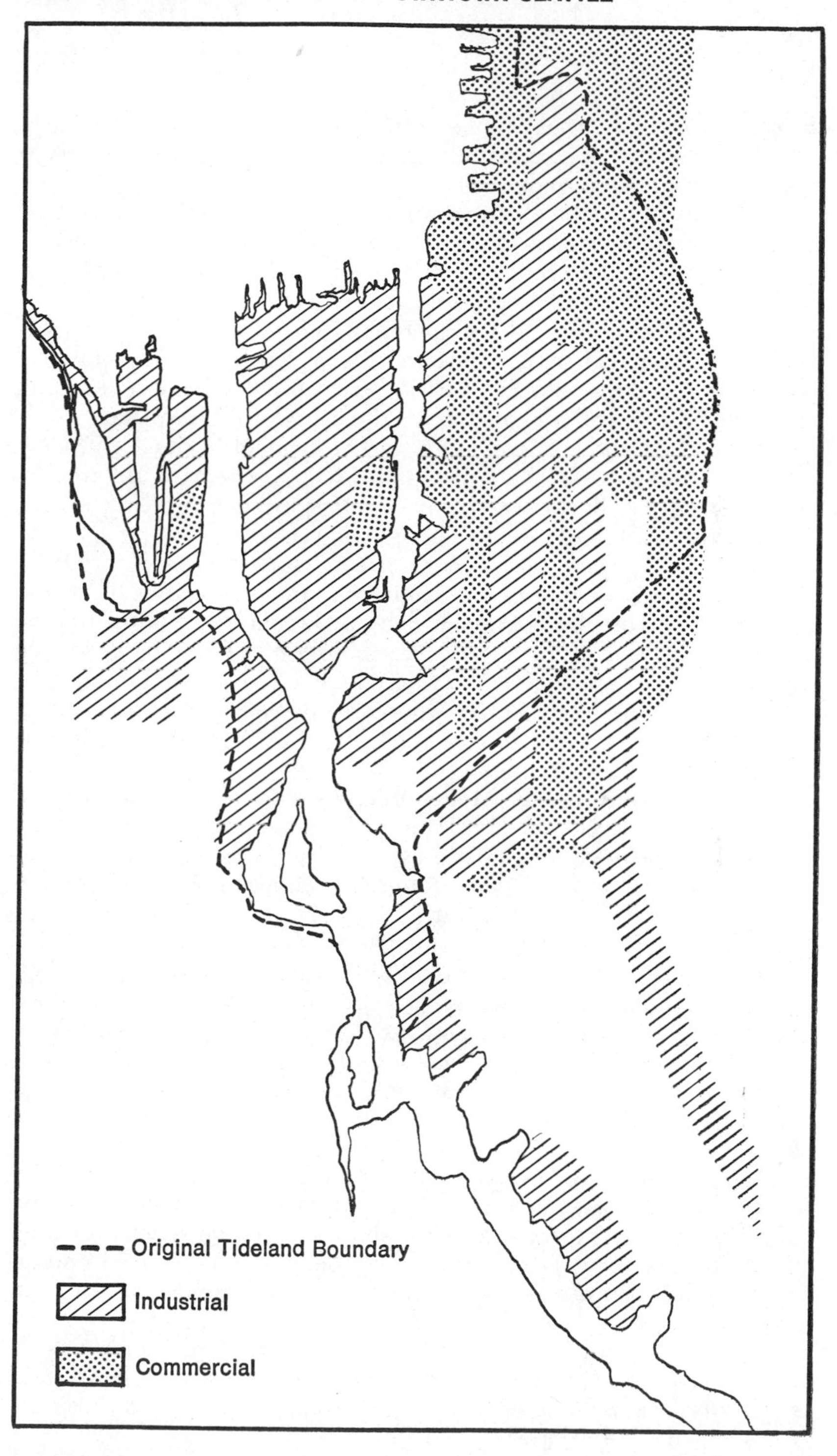

Selected Samples for Further Reading

Blache, J. *L'Homme et la Montagne* (Man and the Mountain). Paris, 1933. A sweeping and somewhat impressionistic geographical survey of man's relations to mountains.

Blanchard, R. *Les Alpes et leur destin.* Paris, 1958. A condensation by the author of his twelve-volume study, primarily of the long history of human use of the French Alps, with a wealth of examples of varying cultural and economic adjustments to a mountain environment.

Eyre, J. D. "Mountain Land Use in Japan," *Geographical Review,* LII (1962), 236–52. Human adjustment to severely limiting physical conditions, and the variety of uses to which an upland area is put, most of them typical of mountain land use patterns elsewhere.

Garnett, A. "Insolation, Topography, and Settlement in the Alps," *Geographical Review,* XXV (1935), 601–17. The effect of differential sunshine, especially between south-facing and north-facing slopes, on the pattern of land use, mainly in Switzerland.

Harrer, H. *Seven Years in Tibet.* Trans. R. Graves. New York, 1954. The fascinating account of an escape from prison camp in India and transit of the Himalayas into Tibet by a German mountain-climbing enthusiast, who also records in this book one of the best existing descriptions of the life of Tibet, where he lived for several years.

Lemert, B. F. "Rural Life in the Mountains between Mexico and Acapulco," *Journal of Geography,* XXXIV (1935), 357–63. A sample study of a mountain community.

Lewis, N. N. "Lebanon—The Mountain and its Terraces," *Geographical Review,* XLIII (1953), 1–14. A study of a mountain landscape and a particular type of mountain land use.

Lunn, A. *The Swiss and Their Mountains.* Chicago, 1963. The influences of the Alps on Swiss culture.

Mason, K. "The Himalaya as a Barrier to Modern Communication," *Geographical Journal,* LXXXVII (1936), 1–16. Self-explanatory.

Peattie, R. *Mountain Geography.* Cambridge, Mass., 1936. The best and most useful single book dealing with the matters considered in this chapter, written in a lively style. It includes detailed discussion of the physical aspects of mountains, as well as of mountain influences on human settlement and society. There are numerous bibliographical references.

Raisz, E., and Henry, J. "An Average Slope Map of Southern New England," *Geographical Review,* XXVII (1937), 467–72. Demonstration of a useful cartographic device for showing relative relief, based on the technique developed by Smith (see below), and especially suitable for small areas, with some discussion of the significance of local relief.

Smith, G. H. "The Relative Relief of Ohio," *Geographical Review,* XXV (1935), 272–84. A discussion of several methods of showing local relief on black-and-white maps, with a sample map of Ohio using a special method developed by the author.

Taylor, J. E. *The Alpine Passes: The Middle Ages.* Oxford, 1930. A fascinating and painstaking historical account.

Toniolo, A. R. "Studies of Depopulation in the Mountains of Italy," *Geographical Review,* XXVII (1937), 473–77. Study of a characteristic trait of mountain settlements—emigration—in an ancient setting.

13
Resources and What They Mean

RESOURCES AND INDUSTRIALIZATION DEFINED

THE WORD "resources" is often taken to mean agricultural and industrial raw materials. It is sometimes implied that soils, water, mineral ores, power, and fuels form the sum of all resources, and also that every lump of coal or drop of oil in the world adds something to the total resource inventory. There is no more reason in this than in saying that food means the seeds of plants and that all plant seeds, including the poisonous or inedible ones, are food; or that literature means poetry and that all doggerel is literature. Confusion is less likely if the word *resources* is given the meaning which general usage and dictionaries give to it: things on which man depends or which support him. Resources have meaning only in reference to man.

But man is not constant, and human changes are accompanied by changes in what constitutes a resource or in what man depends on. This is different at different periods and in different places at the same period. Each civilization rests on a different complex of resources and attaches differing importance to the things which it uses. Modern industrial civilization depends most importantly on power—from mineral fuels, from water, and from fissionable elements. The development of these new sources of power distinguishes the modern industrial revolution from the periods which preceded it, when minerals and many other things on which man still depends were being used.

"Industrialization" may be defined as the use of inanimate power and machines in large units or groups to produce uniform goods in much larger amounts and at a lower unit cost than was earlier possible without power-driven machines. "Industry" is a less precise word which may broadly cover almost any kind of productive enterprise, including agriculture. Ancient or medieval production may also be called industry. The change to what is defined above as industrial or industrialized came perhaps too slowly to be accurately called a revolution. It was spread over at least a century after about 1750, and it is clearly still going on as methods of production change rapidly and new machines, materials, and power sources come into use with what may justly be called a revolutionary speed and a revolutionary effect. Before the 1750's there was a much more gradual refinement and increase in production and techniques, although from the beginning of recorded human history to the eighteenth century there was very much less technological and mechanical change than has taken place in the short time since. These changes were especially accelerated after 1800, and their overwhelming effects on every aspect of human life were and are certainly revolutionary. The phrase "industrial revolution" is in common use, and as long as one understands the time measure involved and the nature of the change, it is a convenient label. Primarily, it reflects an enormous increase in the variety and quantity of resource use. Industrialization brings resources into being not only by discovering them but by making them usable at a bearable cost and with greater efficiency.

Preindustrial Resources

Preindustrial civilizations depended for their support mainly on other things than petroleum, coal, or uranium, but this did not mean that the things they depended on were not resources. Ancient Egypt, for example, used the Nile River as its most important resource, the provider of water and fertile alluvium on which virtually the whole economy rested. It was exploited by means of irrigation; this combination of resources and methods of use did much to shape the character of Egyptian civilization. Phoenicia and Carthage, other ancient Mediterranean states, based their development on their respective strategic locations, exploited by a fleet. Phoenicia occupied roughly the area now known as Lebanon, where the land trade routes from the East reached the Mediterranean. This was a good break-in-bulk point, and it provided the basis for a flourishing trade center. Another local resource was the famous cedars of Lebanon, growing on the mountain slopes of that country near the sea, which supplied timbers for Phoenician ships. Distribution of goods throughout the Mediterranean Basin took the Phoenicians into all parts of the known Western world of that period and even to Britain, but the size, nature, and prosperity of their civilization depended directly on the resource of their location. Carthage is said to have been founded by Phoenician colonists whose quick eye for strategic locations picked out the promontory of Cape Bon jutting into the sea on the African coast opposite Sicily, where the Mediterranean is pinched to its narrowest width and where a city could dominate the trade lanes. The Carthaginian fleet succeeded the Phoenician in the control of this maritime center of ancient Western civilization. Carthage lived on the Mediterranean trade and expressed the completeness of its control by its policy of sinking on sight any non-Carthaginian ship which it found sailing west of Cape Bon.

In more contemporary situations, the basic resource of all pastoral peoples, and even of commercial stock raisers to an extent, is grasslands. Exploited through herds of animals, grasslands are the principal basis of pastoral economies and cultures, affecting their nature as much as the use of the Nile affected Egypt's or as the use of coal and iron affects North American culture. But even modern large-scale mechanized manufacturing may not be wholly or even largely dependent on local deposits of minerals. To take a national example, Japan almost entirely lacks iron ore and petroleum and has only poor and limited domestic supplies of coal. One of Japan's greatest resources is that it is located near the center of the Far Eastern market and served by cheap water transport. This advantage has helped to make Japanese industrialization possible, while in neighboring China, where coal and iron ore are much more plentiful, industrialization was until recently painfully slow and much smaller in scale. A hundred years ago Japan's location and mineral deposits were the same as now, or better, since much of the mineral reserve has been consumed since then, but there was little or no industrialization, nor was there in mineral-rich North America a hundred and fifty years ago. Human knowledge and technical abilities are clearly as important as location or minerals. Location may not be a usable resource, or may indeed tend to isolate an area and retard economic development, until knowledge, techniques, market, and transport have been developed which make a particular location advantageous.

BIOTIC RESOURCES

Natural vegetation and animal life (including fish, birds, and insects) comprise many resources but also include biotic (live) elements which are the reverse of useful and are hindrances to human settlement and economic activity. The forests have been discussed in this connection in Chapter 9, even though forests contain many potentially or actually useful resources in addition to timber. Mosquitoes, flies, lice, rats, and other biotic carriers of disease, or animals, birds, and insects who live by destroying or feeding on the body or works of man, such as gnats, termites, locusts, or crocodiles may in balance be more harmful than helpful to man. Most of the animal and vegetable kingdoms, however, clearly support man as resources rather than making things more difficult for him. Most species are found only in certain climatic areas or in certain parts of the world, and they thus help to make the human occupance and culture of each area distinctive. Hunting, gathering of wild plants and plant products, herding, stock keeping, poultry raising, dairying, field agriculture, and fishing together still occupy and support most of mankind. Over some large areas of the world, in the tundra and the steppe, for example, hunting or herding are the dominant or sole occupations. Without animals human life on any significant scale would be impossible in these areas, and it would be a very different and a very much less productive or rich affair everywhere. Indeed, it seems unlikely that man could ever have progressed from an animal level himself if he had not been able to depend on and to exploit other animals as resources. This is of course even more true of plants, especially since animals also depend on plants.

Man has made relatively little progress in using the vast variety of plant and animal life. Nearly all of our presently useful plants, crops, and animals were used by prehistoric man in one area or another. We have merely spread more widely the most useful plants and animals which were originally confined to small areas, especially the major cereal grains like maize, wheat, or rice, and the common domestic animals such as pigs, chickens, and cattle. We have distributed the products of other plants and animals into areas where they cannot be raised and have refined and made more productive most of the useful plants, crops, and animals inherited from prehistoric man. Compared with the number of animals, birds, and fish domesticated or productively used, however, only two insects, the bee and the silkworm, have been exploited by man directly to produce for him, and only they among the host of insect species can be called resources. Animals are used not only as a source of food or clothing but as motive power in a variety of production and exchange processes, to pull plows, thresh grain, provide power for mills or water lifts, and transport goods. One of the important factors in the relatively low level of economic development in tropical Africa and most of the western hemisphere before the advent of Europeans was the general lack of domesticated animals, except for dogs.

The biotic resources as a whole are the most basic of all, because human life itself would be impossible without them and because they continue to support man more importantly than anything else which he uses. Plants provide directly about 85 per cent of the world's food and indirectly all the rest of it, since all animals and fish depend directly or indirectly on plants. Most fibers are plant material, including many synthetics like rayon, which is made from plant cellulose. Vegetable oils, medicines, dyes, alcohol, building materials, and fuel are only some of the most important of the other plant products which sustain man. Plant and animal products can now more easily be distributed around the earth, and any one area does not itself have to produce them. Because in industrialized economies technological improvements have also made it possible to produce what is needed from plants and animals with only a small part of the total labor force, man may be inadequately aware of his vital dependence on the resources of the biotic world and of the actual bulk of the world economy which their use represents, in terms of area, of numbers of people employed, of goods produced, or of their contribution to our total income and welfare.

The usability of biotic materials depends on the same sorts of factors which govern the use of all resources, however. Knowledge and technical ability are essential, and these differ from place to place and from century to century. Accessibility, distance from market, existing transport means, availability of capital, market preferences and effective demand, and competing sources of equivalent goods also affect the use to which any given biotic or nonbiotic material is put. Fish and fishing, discussed below, make a good illustration of this point.

The Example of Fish

There are lots of good fish in the sea, as we are traditionally told, but only some of them get caught, or, in other words, are used as resources. Those caught for the most part are those which live in or swim into an area accessible to a market which demands fish, or some form of protein, and which cannot satisfy its demands fully or more cheaply from other sources. Not all of the sea has equal amounts of fish, as any fisherman knows. Fish and the materials on which they feed tend to concentrate in the shallower waters above the continental shelves where the sunlight can penetrate, and in areas where cold and warm currents meet or where the water circulates and brings materials up from lower layers or in from other areas. The *continental shelf* is the extension of a major land mass for varying distances beyond the shore line under water less than six hundred feet deep; it may contain shallower areas known as "banks" where the water may be less than a hundred feet deep. The physical map of the world shows the general outlines of the continental shelves and may be compared with a map showing fisheries. Most fish feed largely on other fish, shellfish, water plants, and on minute animal and plant organisms called *plankton,* all of which are especially plentiful in shallower water where currents meet and circulate.

Where such favorable areas lie near major markets which demand fish, the fish are heavily used. The leading fishing nations of the world, in fish caught and men employed, are shown in Figure 23. There are not necessarily more fish off Japan, China, and Korea than anywhere else in the world, but the fish there are close to the world's largest population concentration, one which also has limited domestic supplies of meat or other protein. The same things are true, to a somewhat lesser extent, in northwest Europe and in the Mediterranean Basin and south India, where dense populations with inadequate agricultural production lie close to the sea. There are many more fish in the shallow seas around north Europe (including the Dogger Bank in the North Sea) than in the Mediterranean, but agricultural resources and the means to import food are more limited in the latter area. Proportionately more of the people are engaged in fishing than in most of north Europe even though the total catch is smaller; the fish of the Mediterranean are not

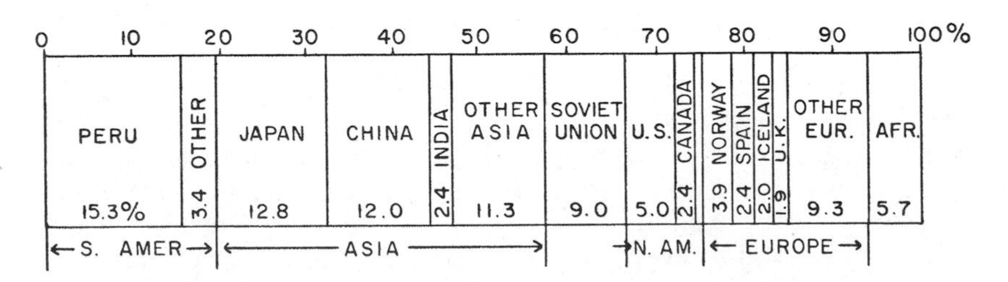

Figure 23. World Fish Catch. Average 1963–1966.

plentiful but are intensively used as a result of economic factors, just as fuel-poor country may make intensive use of wood or of low-grade fuel deposits.

Fishing off the east and west coasts of the United States has a large market to serve, well supplied already with meat and other foods, but at a greater price than fish. The United States also has the major advantage of quick, cheap transport and chilling, canning, or freezing facilities, so that fish moves much farther inland to market than in Asia or even in Europe. South America, Australia, and Africa are less prominent in the fishing business, not so much because there are fewer fish but because the market demand is small, or because the market cannot effectively be reached by existing transport. Most of Africa has a food shortage and protein is particularly scarce; fish are plentiful along parts of the coast but are very little used in the face of the problems of transport, spoilage, and subsistence economic conditions. There are lots of fish off the coast of Argentina, but they are only slightly used because Argentina's economy is based largely on the raising of grains and animals. There is a relatively small population, a large surplus of meat and other foods for export, plentiful domestic supplies at low cost, and little demand for fish. Clearly, what happens to fish in the sea depends importantly on where the nearest land is, what it is like, how many people inhabit it, and what they are doing with it. Location and a variety of economic factors are just as important in determining whether any potential advantage or material will be used as the natural or physical properties of the material itself.

UNUSED "RESOURCES"

If a resource is something on which man depends, then materials, advantages, or locations which for any reason are not usable cannot be called resources. Another example may help to emphasize the variety of factors which affect resource usability and the fact that resources are not simply physical phenomena but come into being or become effective only under certain economic, technological, and other conditions.

Iron Ore in Brazil

There is a huge quantity of rich iron ore in Brazil, in the province called Minas Gerais (General Mines). It is relatively easily and cheaply mined, and there is a great and increasing demand for iron ore on the world market. But this Brazilian ore was until recently relatively little used, either for export or for Brazilian iron and steel production. How is this apparent puzzle to be explained? Consider the matter of exports first. The market for ore lies almost exclusively in the northern hemisphere. Several other excellent sources of iron ore are nearer than Brazil to the major markets in both physical and economic distance: Venezuela, Cuba, Newfoundland, Labrador, and Liberia, among others, not to mention domestic European or United States ores. In nearly all of these places the ore can be shipped to the market more cheaply. In Brazil the major ore deposit lies about three hundred and fifty miles from the sea in rough highland country. The rich Venezuelan ores are much closer to the sea and can be transported on the navigable Orinoco River. In Newfoundland, Cuba, and Liberia the overland distances from the ore deposits to the sea are also small. These ore sources constitute for the American or European market *intervening opportunity*. Brazil has a cost disadvantage, and does not export significant amounts of ore.

Why was Brazilian ore not used domestically to make iron and steel or to support a general industrial complex? To begin with, Brazil has little coal, and indeed South America as a whole is almost entirely lacking in this vital raw material. The nearest major coal deposit is in the extreme southeastern part of Africa, an impractical distance away. Coal suitable for making fair quality coke[1] is present in Brazil, but it is not produced in adequate amounts and is also a long distance from the iron ore in the southern tip of this huge country, an expensive haul for such a bulky commodity. Large-scale heavy industrialization in areas without coal is extremely difficult in competition with other areas which have coal or have economic access to it. Coal is important because manufacturing uses such large amounts of it as fuel, as chemical raw material, or as a source of steam power which may also be used to generate electricity. For many of these uses substitutes are available, but they are at present more expensive than coal except in special circumstances. Electricity, for example, can be used to smelt ores and to make particularly high-quality steel, but at a relatively high cost; the same applies to petroleum or to other fuels. Coal is bulky and relatively low in value by weight, so that it is normally not moved very great distances. Coal also loses much of its weight in being reduced to coke, and most of its weight in the manufacturing process as a whole, when it is transformed into a component of iron, steel, or some other product. Raw materials of which this is true (weight-losing raw materials) are usually not transported very far in their raw state; it is obviously more efficient to use them near where they occur. If an area has large supplies of usable coal, it may draw the other necessary raw materials to it for manufacturing in order to minimize total transport or assembly cost. Such an area may therefore support dense populations and large markets.

Brazil's shortage of coal, important as it is, is not the only factor. For other reasons as well, the Brazilian market was until recently limited. The total population was not large in proportion to Brazil's area. The Amazon Basin, which occupies a large part of the country, is covered with dense rainforest and is very sparsely occupied, mainly by subsistence groups. Most of the rest of Brazil is rugged highlands. All of this contributed to the smallness of the market. Capital or surpluses for industrial investment were also scarce; most of the now rapidly growing development of the iron ores for domestic manufacture was originally financed with United States capital. Small markets and limited or high-cost capital generally mean high production costs, because the economies of scale are lacking. Especially in the heavy industries

---

[1]Coali is reduced to coke (by heating it in an airless oven to drive out the water) for making iron and steel. By no means is all coal suitable for making metallurgical coke. The coal must have the right chemical and structural properties, ash content, heating value, volatility, and other characteristics.

like iron and steel, with their large original investment and fixed costs, the most economic units tend to be large, and, below a certain size, production is too costly to be competitive, either for the local market or for export.

Most of these unfavorable factors in Brazil have now changed. The domestic market is growing in size and in income level, foreign and domestic capital is increasingly available at lower cost, and iron and steel are being made on a significant scale. Other factors may change in future; practicable substitutes for coal may be found, or other technological changes may make it possible for Brazil to support many more people at a higher per capita income level than it now does. The iron ore of Minas Gerais is becoming a resource, however, only as these things happen.

CHANGING MAN, CHANGING RESOURCES, AND HUMAN RESOURCES

Since human civilization began, economic, cultural, and technological changes have continually been altering the pattern and extent of resources. Because such change is not uniform over the world, there are pronounced areal differences at any given time in what constitutes a resource. "Resource" is less a physical fact than a cultural, technical, and economic appraisal. This applies not only to minerals but to all of the things on which an economy may depend, including location, climate, soils, biotic resources, and every usable non-human advantage or material. Japan's location or Great Britain's, California's climate or Cuba's, did not become resources until the technical and economic qualities of their respective occupants and the changed spatial relations of these places with other places made their use possible. Once the non-human resources were put to use population increased, created new human resources, and the income level rose proportionately even more. Non-human resources are part of the physical environment. Man changes or may even eliminate them by using them, and he also changes their influence on him as he increases his own technical powers and as he uses resources more fully, exhausts or destroys them, or pollutes and degrades his environment. Human labor and skills, varying from place to place and from time to time, may also be considered resources and in many situations may be an area's chief resource. This is obvious enough that it should be unnecessary to stress it, but there is a logical and necessary distinction to be made between man on the one hand and his non-human environment on the other. Attention here can be given to the non-human resources as they are brought into being by varying human cultures, technologies, and economies in distinguishing and employing the usable parts of the environment.

POPULATION DENSITIES AND RESOURCE USE

Population densities are understandable only if resources are considered in flexible terms. It has been estimated, for example, that the winter population of interglacial Britain (say in 10,000 B.C.) was about three hundred total. The same area now supports well over fifty-five million people on an enormously higher level of per capita income or consumption, because more of the potential resources of Britain have been put to work by a technically more advanced as well as a more numerous population. In the course of the industrial revolution Europe's population increased seven times, at the same time increasing its per capita income. Europe's latent resources and physical environment changed very little, but they came to be much more effectively used. The same process took place even more strikingly in the United States, whose population increased thirty-one times between 1790 and 1940 as new resources came into being. Obviously, measures of population density or of overpopulation mean nothing unless they can be related to usable resources. Significant measures are not only a matter of the number of people per square mile; the often used term "man-land ratio" may therefore be misleading. The huge area of Baffin Land now supports about three hundred people at a very low standard of living, while tiny Belgium supports about ten million people at a very high standard. The "man-resource ratio" is a clumsier term and may also be misleading unless resources and their use are understood as changeable, but properly understood it is a better measure of the population balance.

Baffin Land is probably no less overpopulated than, for example, China or India with their hundreds of millions of people. China and India will remain overpopulated, in terms of average per capita income or consumption (the best measure of overpopulation), as long as they make inadequate use of their present and latent resources, which in each country are potentially capable of supporting even more people on a very much higher standard of living than now. Europe, and the rest of the world, were also in a position of relatively low living standards and inadequate use of potential resources two hundred years ago as compared with Europe's present position. The West broke out of this situation primarily by means of an economic and technological revolution, first in trade and agriculture and then in manufacturing, which made it possible to use previously neglected resources and to use others more effectively. The effects and implications of the commercial revolution, or the great increase in exchange, beginning especially with the Western Renaissance, have already been discussed in earlier chapters. The agricultural revolution, beginning in the seventeenth and eighteenth centuries and greatly aided by the development of cheap exchange, was no less important than the subsequent revolution in manufacturing. Improvements in the efficiency and output of agriculture made industrialization possible by freeing labor from the farms, by providing a food surplus to feed cities and industrial workers and to form a source of capital for investment, by raising incomes and widening the effective market, and by stimulating interregional exchange. Improvements included the development or refinement of techniques like crop rotation, the use of clover and other leguminous crops, better plowing and fertilization, improved livestock breeds, improvements in the major crop plants and their wider spread, and new, highly productive crops like potatoes (from America) and turnips, which yielded well even in soils or climates which were poor for grains.

In the manufacturing or power revolution which followed, some new resources were put to work, but, just as important, resources which were already known were used much more effectively or productively. Coal, for example, had been used before in Europe and much earlier in China but not on a large scale as coke to make iron and steel, nor in its raw form to produce electric power, steam power, or synthetics. Steel and machines had been made before, but never so cheaply or in such large quantities. New, more complex, and more productive machines were developed, and new sources of power such as hydroelectricity, steam, and liquid fuels came into use; these were in effect new resources. The commercial, agricultural, and industrial revolutions increased the supply of goods so much that the population could not only increase tremendously but could also have a much greater per capita share of goods than before.

POPULATION GROWTH

Where industrialization or commercialization has taken place so far, it has been accompanied by a great increase in population (see Table 1 on page 224.). But after a time, judging from Western experience, the population tends to stop increasing so fast and may even begin to level off or to decline slightly. The reasons for this are not entirely clear, nor is it certain that population necessarily stops growing in the later stages of industrialization. It was expected to do so in the United States by about 1940, but perhaps as a result of the Second World War population increased rapidly in the United States between 1945 and 1970 and has only recently begun to slow down; no one is quite sure why. Industrialization began earlier in Europe and population growth there had slowed down considerably by 1940, apparently reaching equilibrium in some areas; but after the Second World War, European growth rates rose again. Even during the nineteenth century, the demographic experience in different European countries was apparently contradictory and bore no consistent relation to the timing or scale of industrialization or urbanization. In most of the rest of the world commercialization and industrialization are still relatively recent, and populations are still growing rapidly. Not enough is yet understood about population trends and their causes to predict with any accuracy what may happen to population totals or growth rates in any part of the world. Nevertheless, certain basic facts are reasonably clear.

Death rates are high when food and other goods are scarce and when exchange is limited. High death rates and low life expectancies are characteristic of all economies of scarcity; they also retard economic development, since the economy must support people whose productive life is short or who may die before they reach a productive age. The higher proportion of aged people in most industrial societies presents a relatively smaller economic burden. Increasing the supply of goods and the means of exchange, especially for food and medical services, results in a drop in the death rate. This usually happens relatively quickly, although the death rate may continue to fall more slowly for a long time. But the birth rate continues high, usually for several generations, or it may even rise, as it apparently did in the early

TABLE 1

WORLD POPULATION 1650–1970*

(in Millions)

	1650	1700	1750	1800	1850	1900	1950	1960	1970 (U.N. Estimates)
North America	1	1	2	6	26	81	165	199	228
Middle America	6	6	5	10	13	25	51	66	86
South America	6	6	6	9	20	38	111	140	198
Europe (without U.S.S.R.)	96	106	117	156	214	303	392	427	466
Asia	317	362	438	547	793	973	1330	1679	2038
U.S.S.R.	7	17	27	37	60	126	201	214	244
Africa	100	98	95	90	95	120	198	254	352
Oceania	2	2	2	2	2	6	13	16	21
World	535	596	691	857	1223	1672	2461	2995	3633

*In most cases, figures for years before 1900 are necessarily rough estimates, as are some of the figures for years after 1900 for areas without adequate census data, such as Africa, Asia, and Latin America.

Sources: United Nations Demographic Yearbook; Encyclopaedia Britannica; W.S. and E.S. Woytinsky, *World Population and Production* (New York: Twentieth Century, 1953); A.M. Carr-Saunders, *World Population* (Oxford: Oxford University Press, 1936); national censuses; author's estimates from other historical materials.

stages of industrialization in Europe; this results in a large net increase in the population. Economies of scarcity are also usually characterized by high birth rates which are in approximate balance with the high death rates. It apparently takes time for people to adjust to the changed conditions which follow commercialization and industrialization, to lower death rates, and to the fact that all or most of their children will survive to maturity. It is no longer as necessary to concentrate on simple survival and reproduction as it was under the earlier conditions of scarcity. In western Europe it took about two hundred years from the beginning of the industrial revolution for the birth rate to come down into approximate balance with the low death rate and thus to adjust to conditions of abundance. Most of Africa, Asia, and Latin America are still characterized by relatively high birth and death rates, but death rates have begun to fall rapidly and net population growth rates are hence very high. As commercialization and industrialization progress further, the populations of these areas will probably continue to increase at least as rapidly. The hope is that the wider and more efficient use of resources which commercialization and industrialization make possible will be able to keep the supply of goods permanently ahead of the number of people, and that eventually the birth rate will fall. But there can be no assurance of this, and there is growing recognition of the dreadful implications of a world population growing faster than food production. There were more people alive in 1970 than have lived since the beginning of time, a frightening fact. Already the Food and Agriculture Organization of the United Nations estimates that nearly half of the world's present population is seriously undernourished, and that total world food supplies will have to triple by A.D. 2000 merely to maintain the present inadequate nutritional levels. By that time it is estimated that present consumption of energy and of metals will have *quintupled.* For the past several decades world population has been growing by about 2 per cent per year. Such a rate results in a doubling every thirty-five years.

Other Factors in Population Change

There are other, less well-understood reasons for the growth and decline of populations, apart from the effects of disease and war. Not enough is known about the problem even to say for certain whether industrialization is the major factor, although it is clear that populations can increase substantially without industrialization. This happened, for example, in most of Asia as early as the seventeenth and eighteenth centuries, before Western-developed techniques were introduced there. The transition from Neolithic to bronze- and iron-age agriculture must have been followed everywhere by large population increases; there is clear evidence of this in China, India, Mesopotamia, Egypt, and Europe. In much of Asia the great population increases of the past hundred or hundred and fifty years, without substantial industrialization, probably resulted at least in part from increased agricultural productivity, commercialization and exchange, foreign trade, and the introduction of better medicine and public health from the West. In Europe the increase in agricultural productivity which preceded the industrial revolution was accompanied by large population increases. All of these preindustrial or non-industrial increases in population are broadly explainable by increases in the scope and effectiveness of resource use.

Literacy, especially beyond the level of simple ability to read and write, including the interest and ability to read widely and to pursue varied intellectual interests, also seems to be related to declining birth rates. Even in some rural or non-industrial areas, relatively high literacy has been accompanied by relatively low birth rates. In more general terms, when people accept the idea that they can to some extent influence their own economic well-being, they are more likely to limit their families in order to provide better for a smaller number of children. Population may also grow less rapidly in the later stages of industrialization because proportionately more people live in cities. Urban life may offer fewer incentives for large families than life on the farm, and it has so far generally been true that city birth rates in time become lower than rural birth rates. Some students have suggested that urban life may physiologically affect human fertility. On the other hand, France remains mainly an agricultural country, and most of its people live in towns or on farms rather than in cities; yet France's population leveled off about 1910 and subsequently declined slightly. It is also true that in most American cities birth rates have been higher since 1940 than in most rural areas of the United States, although high urban birth rates may be attributable to large groups of industrial workers whose literacy level is low. In most of the so-called underdeveloped world as well, urban birth rates are in many cases higher than rural, although it may be reasonable to assume that this is a relatively short-run phenomenon as traditional village attitudes on the part of recent migrants are replaced by urban ones. But in the meantime, the age structure of the urban population favors high fertility, and infant mortality is usually lower than in rural areas.

These complexities are a good illustration of the handicaps of the social scientist in attempting final or precise answers or prediction based on cause and effect relationships. He cannot usefully test his theories on mice or guinea pigs, he cannot manipulate people, and he can neither successfuly isolate any aspect of human society nor pick out with certainty the particular factors which control the growth of population. He is limited to an analysis of what has actually happened, but for this he often has inadequate information. It is not known, for example, what the exact population of preindustrial Europe was or its standard of living, let alone the populations of the rest of the world before the present century; even now, reasonably accurate census information does not exist for perhaps half of the world. Recent research does suggest that per capita income or consumption, especially food, in early modern preindustrial Europe was much higher than has previously been supposed and significantly higher than in modern China, India, or most of the so-called underdeveloped world of today; it is certainly unwise to assume that Europe's demographic experience will be reproduced elsewhere. There is also much evidence to suggest that birth control in some form, including delayed marriage, was widely practiced and family sizes were limited in many or even in most areas of the world long before the industrial revolution, in response to changing conditions and varying degrees of economic opportunity for the support of population. It is important to remember that although one development precedes or is associated with another the two are not necessarily causally related. A great many factors affect population growth. Broadly speaking, it is known only that population increases have followed industrialization and have also followed other innovations in resource use, and that in time, usually in association with a high degree of literacy and urbanization, birth rates have tended to decline in industrial societies and populations have tended to level off. Both the increases and the leveling off (if it occurs) probably result at least in part from an increase in the supply and exchange of goods, which is in turn the result of a wider and more efficient use of resources.

Malthus and the "Dismal Science"

The British economist Thomas Malthus, whose study of population was first published in 1798, argued that it was the natural tendency for population to increase geometrically, but that the production of food and other goods could increase only arithmetically. Sooner or later, therefore, must come a time when the population exceeds the supply of goods and is thus checked from further growth by famine, or by disease and war, which Malthus realized were related in some degree to famine. He urged social and moral restraint as more desirable controls in order to avoid the less pleasant limits which would otherwise automatically operate if the population became too large. This gloomy analysis helped to earn for economics the title of "the dismal science." Malthus was easy to ridicule because in Europe the publication of his book was followed by an enormous increase in exchange and in industrial and agricultural productivity which destroyed the basis of his argument. Not only were more goods produced per capita, but cheap volume transport facilities made possible the distribution of surpluses and thus virtually eliminated the famines which had formerly resulted from regional crop failures. It is also far from sure that Malthus' analysis of preindustrial Europe was correct, especially in terms of average incomes and of the checks on increases in the population. For much of the rest of the world, however, his analysis may still have some validity, and it can be wholly discarded only when the technically underdeveloped areas undergo the same technological revolution that Europe and North America experiences.

Birth Control and Living Standards

Birth-control programs may not always be effective on a large scale in retarding population growth in situations of economic scarcity. Birth rates are high in such situations, in part because survival is emphasized, and they are encouraged to remain high by high death rates. Various methods of family limitation have been known and practiced since at least the time of the ancient Egyptians. But in most nonindustrial societies children are especially valued as security for old age and as perpetuators of family and individual senses of existence; the childless man or woman is a pitiful figure. Until survival can be assumed, until the economy uses resources effectively enough to guarantee conditions of abundance, until parents can concentrate on providing amply for a few children rather than producing many children in the hope that some will survive, birth control is not likely to be widely practiced even if it is available. This is still to some degree the situation in many so-called underdeveloped economies, although it is gradually changing. In Japan it changed rapidly, after Japan went through a technological and urban-industrial revolution. Economic development is a relative term, and by comparison with the West, Japan is by no means economically or technically retarded. Birth control is widely practiced in Japan, and population has begun to grow less rapidly and may soon level off. The economic margin is wider. Resources are being used more fully, and per capita incomes are much higher than in the rest of monsoon Asia.

Several recently developed techniques of contraception, especially the birth control pill and the intrauterine device (IUD), a simple plastic coil which can be mass produced very cheaply and is reasonably reliable, may represent tremendous positive contributions to human welfare. In much of the world, the impoverishing pressures of population are clear enough to individuals and to planners that a cheap and simple means of controlling conception might be widely adopted. Any projection of current population growth rates produces an alarming result even within the next couple of decades (see Figure 24). Malthus

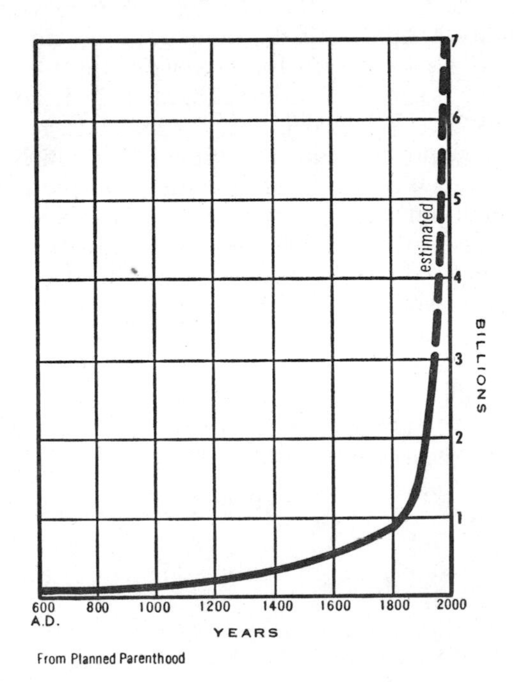

Figure 24. World Population Growth.

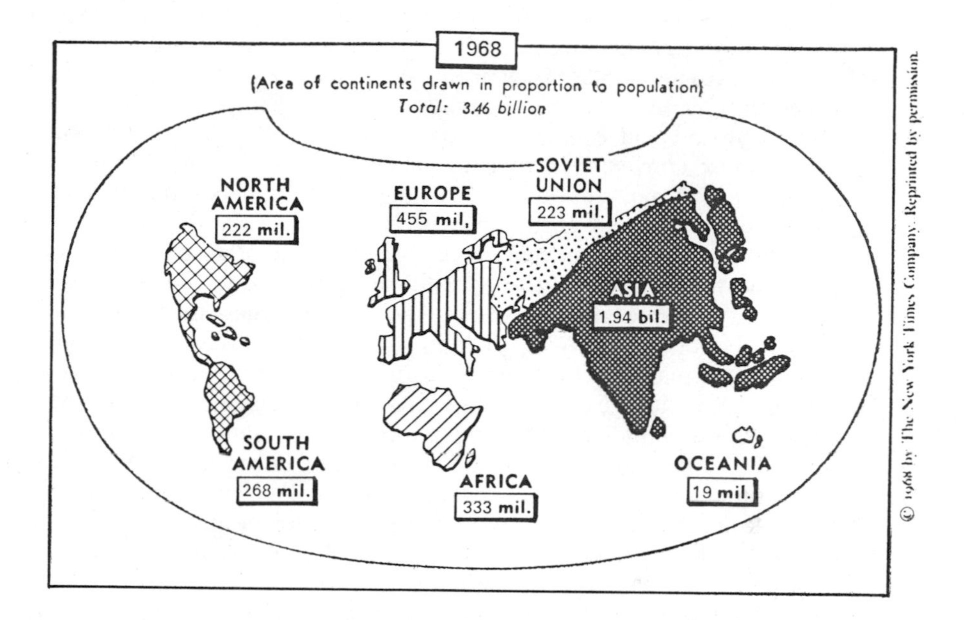

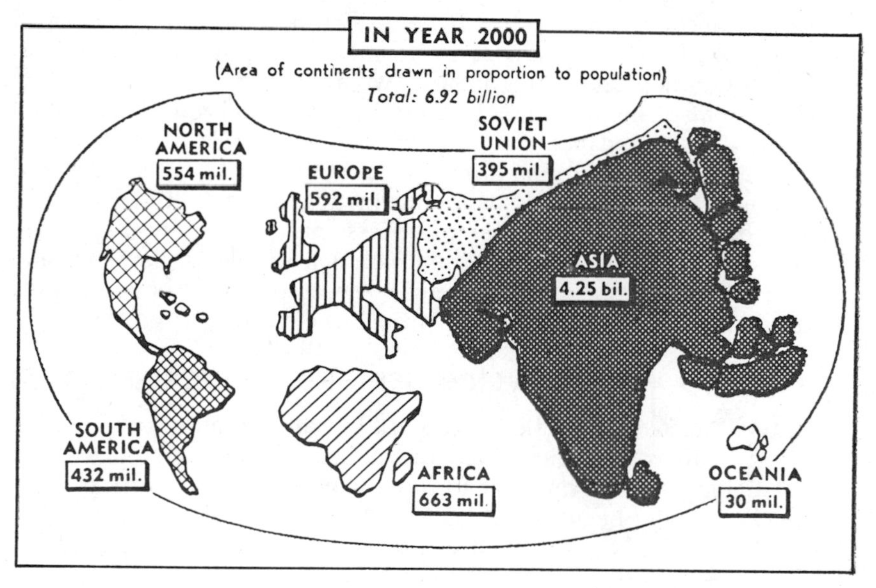

Figure 25. Cartograms of World Population in 1968 and in the Year 2000.

may yet be proved right. If the present perilously low living standards of most of the world are to be preserved, let alone increased, means must be found to control the number as well as to increase the supply of goods before mankind breeds itself into disaster.

Per capita income levels are seldom strictly comparable between different cultural and economic areas, however. They do provide the best and most convenient single measure of economic development, resource use, and population balance, assuming the necessary data are available with sufficient accuracy, which is not always the case.[2] But per capita incomes may conceal major differences in other respects. Different cultures and economies attach differing values to economic goods and to other matters. Subsistence economies may have an acceptable standard of living in their own terms which would seem completely submarginal to individuals in a commercial economy. Such differences may also exist between different subsistence or different commercial economies. Some goods, services, or wants, especially in subsistence economies, are difficult to evaluate in money terms or to convert into per capita incomes. Most Western societies, particularly the United States, believe firmly in the concept of progress and measure it largely in economic terms, or in terms of the number and variety of economic wants which can be developed and can be satisfied: the more realizable wants for goods and services, and the more goods and services consumed, the better. In many or most other societies such attitudes are far less prominent, are wholly absent, or are even contradicted by other and noneconomic values, although societies may acquire some of these essentially Western values and standards as they industrialize. Abundance and scarcity are clearly relative terms, as is economic development, and they will continue to be so even if and when industrialization and commercialization are spread more widely over the world. The process of what is somewhat loosely called economic development involves certain common problems, however, and it is worth examining in those terms. One of the most basic of these common problems is the development of resource use.

RESOURCE USE AND INDUSTRIALIZATION

What makes it possible for potential resources to be used, or what brings resources into being? Why did industrialization take place on a revolutionary scale in much of Europe, North America, and Japan but has only recently begun in China and India, and has still to come in most of the rest of the world? If this question could be answered defin-

---

[2]Completely accurate figures for total income or production are almost never obtainable for any area. Equally important is the fact that income figures may not be comparable. It is difficult, for example, to compare a pound of bananas as income in Costa Rica with a pound of furs as income in Finland. To convert each item into a monetary equivalent conceals real differences in value or desirability in each area. Wants and consumption habits vary widely, for many different reasons. Computation of total or per capita income is usually easiest in commercial economies, where most things are readily measured in money terms, and where relatively more accurate data are available. Such figures are seldom comparable with figures obtained from subsistence economies or from situations where different values predominate.

itively, social science would have come a long way toward living up to its still inappropriate name. Unfortunately the answer can only be guessed at; too little is known and understood about it. Industrialization is still relatively new, and there is broadly only one case from which to generalize, that of Europe-North America, whose development was more joint than separate. Russia and Japan may perhaps provide important newer evidence of a somewhat different kind, as they have applied on a major scale in their own special situations commercial and industrial techniques originally adapted from the earlier Western model. The beginnings of major industrialization in China, India, Brazil, and elsewhere may constitute other examples. But it is hard to say whether any or all of these cases give enough data from which to form adequate conclusions. One can say for certain only that a great number of factors are involved in industrialization and economic development, and that each situation is to some degree unique. The factors include many things besides minerals, soils, climate, or other physical resources. Location, the size and nature of the market, transportation, capital, technology, social structure and cultural values, government policy, the nature, size, and skills of the labor force, managerial and entrepreneurial skills, and nature of economic organization, and popular attitude toward change or toward industrialization must be considered, plus other factors whose relevant importance is difficult to gauge. Presumably the critical factors must also be operative in the right proportions and at the right time. This is a fascinating and important problem to investigate, but it is outside the scope of the treatment here. We are concerned for the moment with what brings potential resources into being and makes them usable.

What Makes a Resource

Human knowledge and technical ability are clearly necessary to make any resource effective. The society must be aware of the resource as something offering support and must know how and must be technically able to use it. The additional factors which make a potential resource usable may be further simplified as the *cost of use* and the *price* offered or *demand* for it. Such an analysis is difficult to apply to resources like location or climate, except as the location of a resource affects the cost of its use, or to some of the things on which preindustrial societies may depend, but it is particularly helpful for minerals, biotic resources, or land. Cost is a measure of human powers, and price or demand is a measure of human wants. Human powers and wants are seldom constant, and their interaction may invalidate resources as well as creating them. A resource comes into being, assuming man knows how to use it, when the cost of its use is met by the price which the economy is willing to offer for it. If cost is greater than price, the material or potential advantage is not usable. This has been the case, for example, with much of Brazil's iron ore. The same may apply to unirrigated land, to remote hydro power sites, to materials which are expensive to refine, or to coal which is buried too deep. The usability of any potential resource also depends not only on the cost of using it but on the cost of using some competitive substitute which may be enough cheaper and may be available in great enough quantity to satisfy the demand entirely. Other iron ore or coal or power sources may be cheaper to use, or other land may be cheaper or more productive to irrigate.

Technology and Changing Resources

The cost factor is continually affected by technological change, which may make it possible to use economically deposits or land or situations which were not resources at all before. The climate of California became a much more valuable resource when man developed low-cost transport links with the American market as a whole, low-cost irrigation, and a variety of other technological innovations. The same kind of analysis applies to the commercial usability of any locational advantages. Changing costs and technology also affect the usability of minerals. Iron and aluminum are found in nearly all soils and are two of the commonest elements in the earth's crust. In this diffuse form they are not now resources because it is so much cheaper to obtain iron and aluminum from the more concentrated deposits provided by nature in certain places. A rich iron ore will contain as much as 65 per cent pure metallic iron; generally, ores of less than 35 or 40 per cent have not until very recently been economically usable except in special situations. But suppose that someone developed a powerful magnetic extractor which could be passed over the ground like a vacuum cleaner picking up the iron. If the costs were low, this bit of iron in the soil might become a resource, provided costs for extracting and transporting iron from the existing ore deposits did not fall even lower.

The industrial revolution has depended on technological developments essentially like this example. The cost of obtaining and using raw materials and power has progressively been reduced so that a great variety of latent resources have become usable, including land which could not economically or productively be cultivated when technology was less advanced and costs of use were higher in proportion to the yield. The transformation of coal into power (rather than burning it only for heat as before), through the medium of the steam engine, one of the first modern technological developments, is an example of how industrialization has also changed and multiplied the usefulness of pre-existing resources. Extraction and transport costs for coal were also greatly reduced with the help of machinery and inanimate power, and coal became a much more important resource than it had been before. Aluminum has become a resource only recently. Technological improvements in extraction and refining have reduced the cost of aluminum by nearly three thousand times since 1860; with each cheapening it became usable for more and more purposes. Originally aluminum was used only for small, high-value goods like cooking utensils; now it may be used extensively for structural purposes or for bulky transportation equipment. Steel went through the same kind of process a hundred years ago, when the progressive lowering of costs made possible by improved technology made steel the cheapest suitable metal for most purposes. Aluminum has many superior physical properties, including resistance to corrosion and great strength in proportion to its weight, but it cannot displace steel for many of the uses where it would physically be suitable as long as it is more expensive. Copper also has many desirable properties and would be preferable to both steel and aluminum in many cases, but the higher cost of copper at present confines it to uses where its special properties are particularly needed, such as for conducting electricity. Many metals are interchangeable for a variety of uses; relative costs mainly determine which metal is used. Relative costs may change, and may thus affect resource use and resource appraisal. Copper and aluminum are particularly close competitors for a number of uses, so that slight changes in relative costs can have substantial effects on the consumption of both. One development which helped to make steel cheap was the perfection of low-cost techniques to use lower-grade ores or ores containing impurities (especially phosphorus) which were unusable before. This also had the effect of greatly increasing the total reserves of iron ore and fundamentally changing the resource inventory. France, for example, found itself, as a result of this discovery, well endowed with iron ore, whereas earlier usable French ores were limited because of high phosphorus content.

As a resource continues to be used, the criteria applied to it may change. Copper, for example, has been mined in the United States long enough to exhaust the richest deposits. Fifty years ago material containing less than 4 per cent pure copper was discarded as worthless; now material containing only 1 per cent copper is considered usable ore, and in some cases it pays to work over the discard piles of fifty years ago. Technological change is largely responsible for this, since it has so greatly reduced the costs of extraction and refining. But in the Congo, where rich copper ores have only recently begun to be mined, 4 per cent is still a basic acceptable minimum. The original deposits of rich iron ore in the Mesabi Range near the western end of Lake Superior are approaching exhaustion. They have been the major source of iron ore for American manufacturing not only because of their high quality but because of their fortunate location close to the water transport network of the Great Lakes, even though they are not very close to the major markets in real distance. As these ores approach exhaustion, new steel plants have been built on the east coast, where ores can be brought in from overseas by cheap water transport. But there are also extensive deposits in Minnesota of a rock called *taconite* which is 25 to 35 per cent iron. Until recently it has been too expensive to get the iron out of taconite, but a process of extraction and refining or concentration has now been developed which can produce iron from taconite at a cost competitive with other sources of iron. Taconite has become an ore and has thus become a resource.

Changes in Demand

Technological change also affects price or demand, since it changes the market's ability to use things and may create or remove needs for certain goods. There is no longer a demand for whalebone for corset stays, and the demand for horseshoes, goose grease, woad, or wolfbane has greatly decreased, not only because cultural preferences have changed but also because technological change has brought about substitute demands or has developed alternative goods or sources which are cheaper or more effective. Totally new demands may also arise. A hundred years ago there was little demand for rubber except as erasers—this was the origin of the name *rubber*. The development of the automobile in particular has created a tremendous market appetite for rubber, as industrialization has created a great variety of new demands for other new resources. A hundred years ago petroleum had almost no uses except as medicine; electric power was just growing out of the toy stage and had few uses. Technological change reduced the cost of obtaining and transporting the new resources, but it was also responsible for bringing about the demand for them in the first place, or, in other words, for bringing the resources into being. As demand increased, efforts were further concentrated on increasing production and lowering costs. It was the interaction of increasing and expanding demand and falling costs which created new resources, as in other cases declining demand removed items like whalebone or wolfbane from the resource list.

When real or artificial shortages arise, as for example in wartime, price rises or increases in demand may also make resources out of land, minerals, commodities, or strategic locations which were not resources before, or whose cost kept them from being usable when the price or demand was lower. Mines abandoned because the richest deposits were exhausted may be reopened if ore prices rise or if the minimum acceptable standards for ore fall. Abandoned or marginal land may be cultivated. Normally uneconomic substitutes for various kinds may be pressed into use. In normal times areas which lack coal may develop hydro power sites which would not be developed in coal-rich areas. Areas which are poorly endowed with mineral resources in general may use deposits or materials which would not be considered resources elsewhere. Demand may thus create or eliminate resources by affecting the resource appraisal. But technological change tends primarily to reduce costs and to widen demand rather than to raise prices, and it is in this way that most new resources come into use. Many more resources are now used than before the industrial revolution, including many which had little value or were even unknown a hundred years ago. This is usually a cumulative process, for very few resources in the general sense, as opposed to specific deposits or individual sources, pass out of use entirely. Man still makes use of nearly all of the things used in the distant and more recent past, although he may attach much less relative importance to certain things. Each year new resources are added to the list, and it is difficult to predict what may be included in the future. The rate of technological change is still accelerating.

WHAT MAN DEPENDS ON

The above discussion of land, minerals, or other commodities should not draw attention away from the broader and valid concept of resources as it was defined earlier: things on which man depends. Climate or soil may be a more important resource in this sense than coal, as may location or animals or fresh water. Sea water may become a major resource if the variety of minerals which it contains can be extracted from it cheaply enough to meet or beat the cost of present sources. Resources like climate or location are, however, difficult to evaluate completely in precise cost terms. Technological change affects the usability of climate or location as it affects other resources; they become more or less usable in accordance with changes in man's powers and wants, as indicated by the examples of California's climate or Japan's location. Two of the leading manufacturing industries of California, airplanes and films, depended in part, especially in their early growth, on the resource of the local climate, as Japan's economic development is dependent in part on the resource of her location. Such resources are directly relevant to industrialization, but since their effects are less easily reduced to detailed measures, it may be more helpful to analyze selected samples of resources which can be so treated and which will illustrate the general problem of resource usability.

Resources and the Location of Manufacturing

Of the many thousand raw materials on which an industrial economy is dependent, a few are of paramount importance, especially as they help to determine where industrialization may occur. These few are important because modern manufacturing uses very large amounts of them and therefore cannot normally afford to haul them from great distances. The mineral alloys which are necessary to make steel hard or resistant to heat or corrosion (see Table 2) are just as necessary as iron ore for the making of that steel. But such relatively small amounts of these alloys are needed for every ton of steel that they can economically be brought from the other side of the world if necessary. Some things which are used on a large scale occur or can be produced only in parts of the world where there are no other advantages for manufacturing; they therefore do move considerable distances to reach the established manufacturing centers. High-value tropical crops like rubber, chocolate, or vegetable oils do this. Mineral deposits occur in widely scattered locations, and only by chance do they coincide with favorable climate, landforms, population centers, or other factors conducive to manufacturing. When they do not, they may not be usable, if they are bulky or heavy in proportion to their value.

There are only a limited number of major heavy manufacturing areas in the world: northwest Europe, the Ukraine-Urals area in the Soviet Union, northeast United States, central and western Japan, and north China-Manchuria, and the Damodar region of India, but there are plenty of mineral deposits distributed around in other places. Heavy manufacturing, in which raw materials and transport costs are a large proportion of total production costs, must concentrate where assembly costs for its raw materials and distribution costs for its finished products to market are at their lowest. It must be particularly sensitive to its economic distances from the raw materials of which it uses the largest amounts. If these economic distances are small, and if distances to market are also small, heavy manufacturing can ignore distances for other materials which it needs in lesser amounts. Large deposits of copper, aluminum, manganese, antimony, tungsten, or nickel are unlikely by themselves to attract much manufacturing locally, aside from the immediate processing of the ores, because they are used in relatively small total amounts in manufacturing as a whole. If an ore contains only a small proportion of pure metal, it may not pay to ship long distances what is mostly waste material. Such ore is usually somewhat refined on the spot. Copper ore, for example, is often smelted, or concentrated into *blister* copper, at the mines; what is sent to market or to copper-using industries from smelters is about 99 per cent pure copper. But heavy or general manufacturing would find little advantage at a copper or tungsten mine if the mine were far from sources of fuel and power, from other bulky raw materials, from labor, and from the market.

Heavy Manufacturing and its Role

Heavy manufacturing, which includes most importantly iron and steel, construction materials, transport equipment, chemicals and refining, and heavy machinery, is fundamental to industrialization because it provides the essential means of producing and exchanging more goods by power-driven machinery. Without these basic essentials, or *capital goods* (sometimes called *producers' goods*), it is difficult or impossible to produce or exchange *consumer goods* like textiles, furniture, or food for wider markets, except in small quantities or at a relatively high cost. Thus all industrialization in the modern sense depends on the heavy industries. The supply and cheapness of fresh and canned foods, ready-made clothing, houses, or beds depend upon steel, machinery, and transport equipment. Consumer-goods industries, most of which are mainly market-oriented, since raw materials are for them a relatively small proportion of total production costs, are often attracted to centers of heavy manufacturing and their attendant markets, power, services, capital, labor, and transport facilities. Consumer goods are made nearly everywhere in some form and to some degree and have been for thousands of years. But they are made cheaply and in large quantities only where they can depend on heavy industries to supply them with the means of mechanical production and exchange.

The Big Three: Coal, Iron, and Oil

Coal, iron ore, and (more recently) petroleum form a triad of heavily used resources to which most manufacturing must be especially sensitive. Fresh water might well be included, since it is used in even larger amounts by most manufacturing, but fresh water is much more widely available and for the most part influences the location of manufacturing on a smaller or more local scale.

1. *Coal.* Of the big three, coal was long the most important, because it was the cheapest available material for a great variety of uses and was therefore consumed in huge quantities. As pointed out earlier, coal is often important enough in this sense to draw to it whatever other raw materials are needed and thus to mimimize total transport or assembly costs, if the factors of location, market, technology, capital, labor, and transport are favorable enough in the area where the coal occurs. There are very few large industrial areas which are not

near to coal, and most of the largest ones are very close to coal, at least in economic distance. Coal is used not only in the metallurgical industries as coke (see note 1 on page 220), but as the largest single source of electric power (from steam plants, where power is generated by steam-driven turbines), as a leading raw material in a number of chemical industries, where it is especially important for production of synthetic substances such as dyes and plastics, as a fuel for domestic heating, and for many other purposes. As petroleum, hydroelectricity, and fissionable materials have become cheaper and increasingly important sources of power, coal has lost some of its former proportional dominance. For example, coal is no longer the chief fuel for domestic heating or for railways or ships. But for many years coal will remain the best or cheapest material for a variety of manufacturing needs, and it continues to be the most used industrial raw material, after fresh water.

It is possible that coal's relative importance may increase in future, at least for a time, since world petroleum supplies will almost certainly be exhausted long before world coal deposits. Oil can be obtained from oil shales, which are relatively plentiful, but at what may in future be a higher cost than oil from coal. Petroleum or its products can be made from coal, as is done already on a small scale in certain areas where petroleum is scarce, such as in Germany during the Second World War. The coal-rich areas will probably continue to enjoy major economic advantages, as they have since the eighteenth century. Much of the enormous and rapid industrial development of the United States was due to the fact that it is "the land of cheap coal," with about half of the world's total workable coal reserves, plus large shares of the world's petroleum, iron ore, and other valuable minerals. Extraction costs for coal are also relatively low in the United States, due to the favorable nature of most of the deposits. But the United States is rapidly consuming its minerals and is already a large importer. With the fortunate exception of coal, we may shortly be a net importer of most of the heavily used minerals. American use rates are many times the use rates in the rest of the world and are still rising.

2. *Iron Ore.* Since iron ore is used in somewhat smaller total amounts than coal, it therefore is usually transported to places where coal is cheap. But manufacturing uses enough iron ore to have its location profoundly affected by ore availability, as illustrated by the trend of migration of the iron and steel manufacturing in the United States toward the east coast in response to changed supplies of iron ore. Technical economies in the use of fuel have decreased the proportional amount of coal necessary for iron and steel making (and for other manufacturing processes also) so that slightly more iron ore than coal by weight is now used in making a ton of pig iron. But coal has many other uses, and steel manufacturing also makes large use of scrap metal as a raw material, mixing it with pig iron to make steel and to some extent replacing iron ore. Scrap is usually cheapest and most plentiful in the large market centers, which are often associated with coal, import coal, and consume large amounts of metal products. Thus coal's locational dominance tends to remain, as well as its association with the dense populations necessary for manufacturing, supported by man-

ufacturing, and constituting large markets. Comparison of the distribution of coal deposits and population in Europe provides a suggestive illustration.

Markets, the transport facilities which serve them, and their pools of labor, capital, special services (such as engineering, banking, or insurance), and consumers are a major attraction for nearly every kind of manufacturing, wherever transport is cheap and adequate enough to bring the necessary raw materials to the market at a bearable cost. Coal is closely associated with markets and dense populations in so many cases because coal is too bulky and low in value to be transported far in the huge quantities needed, and the markets and populations have thus frequently grown up close to the coal. As transport facilities improve and transport costs for bulky raw materials decrease, the market becomes an even more desirable or profitable location for manufacturing. This process has continued throughout the industrial revolution so far, as the sites of raw-material and power sources have increasingly lost their locational attraction and the market has gained. Since so many of the markets grew up in close association with coal, and since coal is still used in such large amounts, this development has not affected coal as much as it has affected other raw materials.

Where water transport is available, iron ore may move considerable distances to market centers where coal, scrap, other necessary raw materials and power, labor, capital, and transport facilities are present. Specialized carriers also help to make the movement of iron ore relatively cheap despite its bulk and weight.

3. *Oil.* Petroleum is much more foot-loose than coal or iron ore, even though it has also come to be used in huge quantities. Petroleum is much more cheaply transportable, by pipeline overland (almost as cheap as ocean transport) and by tanker at sea. It can be loaded and unloaded quickly, entirely mechanically, and thus cheaply; special carriers built to haul nothing but petroleum and loaded to capacity with no waste space can offer very low haulage rates. The relatively low transport cost for petroleum is fortunate, since so much of the world's petroleum lies in the Middle East, the East Indies, Mexico, and Venezuela, far from major market centers. Europe has much less petroleum, but this has not been a strong deterrent to industrialization in the past, though it is strategically awkward in wartime. Europe as a whole is still the largest single industrial area of all, and it can economically draw petroleum to itself by sea or by pipeline. A modern industrial area must now have low-cost access to petroleum, as Europe has, even though petroleum alone does not create industrialization where the deposits occur. Oil can too easily be moved to places which have greater total advantages for manufacturing. The oil of Iraq or Saudi Arabia, for example, is not necessarily a stimulus to local industrialization except as it can provide capital which may be invested in manufacturing; other requisites for industrialization, such as other raw materials, technical skills, adequate political leadership, market, and skilled labor are largely lacking, and oil alone cannot attract or create them.

WHAT MAKES A MINERAL DEPOSIT USABLE

Mineral resources are only part of the resource concept. They are a distinctive part because, unlike climate or water power, they can be exhausted by use and are not renewable; this imposes more rigid economic conditions on their exploitation. Mineral ores and fuels are the most important members of this group; the conditions of their exploitation can be taken as typical of other minerals, and in a broader sense of all resources. What helps to make a given mineral deposit usable as a resource? The factors include the following:

Physical Factors

1. *The size of the individual seam, bed, or pool.* Particularly with mining, this will be reflected in costs. If the coal seam is thin, or bent and broken, it will cost much more to get the coal out than if it occurs in a thick regular layer. If the seam is too small to permit mechanical mining, costs rise steeply. This has happened in many of the older coal mines in Great Britain and Japan, where the thickest seams have been worked out. The United States is still skimming the cream of its coal deposits, and costs are correspondingly lower. The same considerations apply to ores and to a degree to petroleum.

2. *The depth below the surface at which the material lies.* The relation of this to the cost of extraction is obvious, though it is often ignored by "resource" surveys. What lies on top (the overburden, as it is called) must either be removed or deep shafts sunk, and thus heavy costs are undergone before there is any yield. Inadequate size or excessive depth are largely responsible for the fact that more ores and coal must be left in the ground than can economically be brought to the surface, even where a mine has been dug. If present experiments with burning coal in place underground can produce an economic source of power or by-products, the resource status of nearly all coal deposits would be profoundly affected.

3. *The quality of the material.* Standards for this may change, as discussed above, but in any given situation the grade of the ore, coal, or petroleum will help to determine whether the costs of extraction can be repaid. Naturally, a high-grade deposit will make more expensive extraction worthwhile.

4. *The size of the total workable deposit.* This also frequently leads resource surveys astray. For modern industrial operations, a deposit must be large enough to enjoy the economies of scale. To act as a localizing factor, it must also be big enough to support the industrial operation for a long time. A steel plant, for example, cannot afford to pick its location on the basis of an ore deposit which it would exhaust in two years. In practice, this factor rules out a great many deposits for which the other factors are favorable or which were used earlier when the scale of manufacturing was smaller. Small bodies of iron ore in New Jersey and Massachusetts, for example, which once supported a flourishing industry in the eighteenth century, using charcoal as fuel, are now of only academic interest, mainly because of their limited size.

This is typical of similar small bodies of ore in a great many places elsewhere, many of them once smelted with local charcoal but now obsolete even though they may be of high grade and have low extraction costs. Improvements in transportation, widening of the market, the change to dependence on coal rather than on charcoal, and the greatly increased size of manufacturing units with attendant lower costs of production have resulted in a concentration of production at a few sites where very large supplies of ore and coal are cheaply available. State or national totals of mineral sources may thus mean very little.

5. *The distance from the market or from the manufacturing site.* This will mean primarily economic distance, reflecting topography, other aspects of the intervening space, and available transport facilities as well as physical distance.

Economic Factors

1. *The size and nature of the market, or the effective demand.* The market must be willing and able to pay for raw materials and to use them profitably. Most preindustrial economies cannot do either on a large scale; hence their coal or ores may not be resources for them.

2. *The availability and cost of capital.* This is primarily a reflection of per capita incomes or of the size of the economic margin. Plentiful capital at low interest rates is essential for industrialization and for the use of industrial resources. The economy must be able to save, or to set aside surplus goods and postpone consumption, in order to meet the heavy costs of building a factory or a railroad; these capital goods, or producer's goods, cannot be eaten or worn but are used to make and exchange goods which can be consumed. The increases in the production and exchange of consumer goods which the production of capital goods makes possible do not take place immediately, and there may be a wait of several years before the results are available to the consumer. Without ample surpluses or capital for investment, this postponement of consumption and its later rewards is not possible; industrialization cannot take place, coal mines cannot be dug, and latent resources cannot be used. If investment capital is expensive, as it nearly always is wherever it is scarce, resource use and industrialization are retarded.

3. *The transport media.* Without low-cost volume carriers, bulky goods cannot economically be moved, and many materials cannot be used.

4. *The technological means available.* These have already been discussed.

All of the above factors help to determine in any given case whether a particular material is usable and whether cost and price or demand are in appropriate balance. The weight of these factors differs at different times or in different places. Japan, with a domestic coal shortage and few rich deposits left to draw on, mines coal which would be ignored in the United States. But China, with much good-quality coal, has until recently used relatively little of it, either because its location was wrong with respect to the market, or because the market and the supply of capital, transport, and technology were unfavorable.

The Japanese market calls forth what in other situations would be impossibly high-cost coal because there are no cheaper alternatives available, and makes proportionately large use of oil to generate electric power, since oil can be shipped more cheaply than coal. Domestic resource inventories also help to account for the development of hydroelectric power in coal-poor Italy and France as well as in Japan. Thus resource appraisals differ with differing economic as well as technical and cultural conditions.

In Chapter 9 and Chapter 12 the same kind of analysis was applied to timber as a resource. There the economic distance to market is particularly important because timber is so bulky and heavy in proportion to its value and therefore cannot bear expensive transport hauls. Problems of mining are roughly paralleled by the problems of getting trees from steep slopes in remote mountains or out of the barrier country of the rainforest, where mobility is limited and where there are local physical obstacles contributing to high extraction costs. Thus many forest stands are not resources because they are not economically usable; cost and price are not in appropriate balance. Chapter 10 analyzed soils as resources in the same way. The set of factors outlined above helps to determine the usability of all mineral deposits, including the many nonmetallic ones like salt, building stone, sand, gypsum, nitrates, or sulphur and also affects potential resources which are less concrete, like climate or location. Changing and widening uses of climatic and locational resources have already been referred to above and in earlier chapters, including the examples of California in 1492 and the present, Japan over the past hundred years, and western Europe or Britain since the Middle Ages. Fish were discussed in an earlier section of the present chapter as an example of the biotic resources. Hydroelectric power as a resource has also been discussed, in Chapter 11, but it may be useful in summary of the resource concept to examine hydro power in more detail by applying to it the same kind of analysis used above for mineral deposits.

WHAT MAKES HYDRO POWER USABLE

Whether a given site is usable to produce hydro power depends for the most part on the following factors (assuming that capital and technology are adequate):

1. *Whether the stream can be easily or cheaply dammed or diverted.* This will depend not only on the contour of the land but also on the strength and imperviousness of the local rock.

2. *Whether there is an adequate volume or force of falling water at the site which is constant enough throughout the year to permit economic operation.* If the stream loses most of its volume or dries up for part of the year, it may not pay to invest the large sums necessary to make use of the balance of the year's flow. This is the case with many sites, especially in climates where there are marked seasonal differences. The minimum flow at mean low water, or the *firm power base*, is also important, since it may be too small, even if it is relatively con-

stant, to make a dam pay, although the possibilities will vary depending on how much water storage is possible. Here the economies of scale operate, as with mineral deposits.

3. *Whether there is a high constant load factor within a three-hundred-mile range which can make the heavy investment profitable.* As pointed out earlier, electric power cannot be stored, but the fixed costs of hydro power installations are high. Ideally, the load factor should include a large proportion of industrial use; manufacturing is a much bigger and less fluctuating consumer of electricity than households. In most market areas there is a series of peaks and hollows through the day and night in the consumption of power, reflecting rush hours, meal times, and the hours after midnight when most people are asleep. Manufacturing can help even out these irregularities in demand, but usually a city will also have to maintain power plants producing from coal or oil (steam-driven turbines). In most big cities, as already pointed out, coal and oil are still the largest sources of electric power.

4. *The other benefits whose value could be charged against the cost of the dam*: navigation, flood control, irrigation, and so forth. Whether these uses are possible depends on the nature of the site and of the areas around it.

Hydro power is clearly not a "free" resource as it is sometimes called. Even in the cases where it pays to develop it, where all the above factors are favorable, the heavy capital investment necessary may still mean that the hydro is a more expensive source of power than coal or oil, or at least that it cannot entirely replace them. It is "free" only in the sense that it is renewable, like soil or climate, whereas mineral deposits eventually are worked out and, as their exploitation increases, are likely to become more expensive. But whatever the long-term, strategic, or even present virtues of hydro power, it costs money; it becomes a resource only when this cost is low enough to meet the market's price in competition with other sources of power.

RESOURCE CRISIS OR A BRAVE NEW WORLD?

There is a continuing public and private debate between those who are alarmed at the rapid depletion of mineral, water, biotic, and soil resources and those who argue that the equally rapid progress of technology will create new resources by finding new sources of power and materials and by decreasing the cost of using the present ones. One side stresses the undeniable fact that world population is still increasing rapidly and will presumably increase even more rapidly as industrialization spreads (see Figure 24). Good agricultural land is limited, and possibilities for expanding profitable cultivation may not be great enough to compensate for the larger number of mouths to feed. While only about 10 per cent of the earth's land surface is now cultivated, the remaining 90 per cent (including the deserts and ice-caps) is neglected for good reasons. Any major extension of cultivation is likely to involve the use of marginal land. This not only presents cost problems but means that, whatever the cost, such land is not likely to be as productive as the existing cultivated land. Rising market demand and prices may make it profitable to cultivate new land, but the usable supply is restricted, and the returns diminish as cultivation is extended onto less and less good land. Mineral ores and fuels are rapidly being used up. The world consumed more minerals in the first half of the twentieth century than in all previous history; the rate of use is still rising and will continue to rise, particularly as more of the world becomes industrialized. Half of all the coal ever mined has been dug since 1935, half of all oil since 1958! These resources, which took 100,000 years or more to form, will have been largely depleted in about a hundred years of large-scale industrialization. It is argued that technology cannot increase resources or production in any sphere fast enough to keep pace with population and consumption increases, especially when it is confronted with finite or disappearing supplies of land and raw material.

This is essentially the same dilemma which Malthus described for preindustrial Europe and which he mistakenly assumed would continue in the nineteenth century. The "Neo-Malthusians" argue that because the West escaped temporarily from the unfortunate consequences of that dilemma is no guarantee that the West, and more especially the rest of the world, can continue to escape in future as world population growth gets into its stride and available resources are depleted. Some estimates suggest that tin, lead, and zinc, in terms of their currently classified ores, may be largely exhausted by the end of the present century at current or expected rates of use; so may commercial timber by present standards, and iron ore may very shortly follow. Deposits of coking coal and petroleum which are now considered economically usable may all have been consumed in another century.

The reply is the truism that resource appraisals are not constant; they change as man changes. At least so far, as rich ores, deposits, soils, or forests have been exhausted or taken up, stronger demand, higher prices, or lower-cost and more efficient techniques of use, or a combination of all of these things, have often made it possible to use leaner ores, lower-grade and more remote deposits or forests, oil shales, or poorer land without necessarily decreasing average yields or per capita consumption. Better recovery techniques may also make it possible to get more of the oil, coal, or ore out of a given deposit; at present, especially with coal and oil, more is left buried than can be taken out, although proportionately more and more oil is being recovered by pumping and other methods. New deposits are also almost certain to be discovered. Even in relatively well-surveyed areas like the United States, new mineral concentrations are continually being found. Admittedly, this cannot go on indefinitely. Total world coal reserves may last for another two thousand years, but for coking coal and most other minerals, especially the metals and petroleum, the end may already be in sight in terms of present techniques. Equally alarming is the galloping rate at which man is destroying vital resources of air and water through pollution.

The crux of the matter is that techniques change and are now changing particularly rapidly; this means in effect a change in resources. Yields of all resources are more likely to increase than to decrease, judging from past experience, although there can be no assurance that yields can incease as much or as fast as population or consumption, if present growth rates remain unchecked, or that pollution can be halted before the environment is irreparably damaged. Accessible sources of charcoal, for example, were largely exhausted in England and elsewhere by the eighteenth century; the metallurgical industries of that time were rescued from what appeared to be a major and crippling resource crisis by the development of coal and coke as metallurgical fuels. The same would apply to the shortage of sizable oaks for ships of the line and the rescue of the British navy by the development of iron and steel ships. If and when the bottom of the barrel is reached in terms of presently known and appraised materials, as it may well be in the not very distant future, the hope is that technology will have progressed enough to extract at a bearable cost whatever minerals are needed from almost any rocks or soil or from sea water. Fuels and power may be derived from nuclear fission or from the sun. Pollution can be controlled by already known technology. The problem here is political, economic, and social rather than technical. Food and plant cellulose may be produced by chemical synthesis, by artificially manipulated and accelerated growth of certain prolific plants like algae and yeast, or by planned cultivation of the oceans and the continental shelves, as well as by a more productive use of the land and of fish, animal, insect, and bird resources. Some enthusiasts argue that in another two or three centuries the world might support twenty or thirty billion people at a much higher average income or consumption level than the present world population of about three and a half billion enjoys. But there remains the fundamental and disturbing question of whether the production of food and other goods can in fact be increased fast enough and at a bearable cost to keep pace with galloping population increases, especially in the so-called underdeveloped world. The two sides of this controversy are amusingly pictured in the following impromptu lines, written in 1955 by Professor Kenneth Boulding while he was attending a conference on "Man's Role in Changing the Face of the Earth":

A Conservationist's Lament

The world is finite, resources scarce.
Things are bad and will be worse.
Coal is burned and gas exploded,
Forests cut and soil eroded.
Wells are dry, and air's polluted,
Dust is blowing, trees uprooted.
Oil is going, ores depleted,
Drains receive what is excreted.
Land is sinking, seas are rising—
Man is far too enterprising.
Fire will rage with man to fan it
Soon we'll have a plundered planet.

People breed like fertile rabbits
People have disgusting habits.
Moral:
The evolutionary plan
Went astray by evolving man.

The Technologist's Reply

Man's potential is quite terrific;
You can't go back to the Neolithic.
The cream is there for us to skim it—
Knowledge is power, the sky's the limit.
Every mouth has hands to feed it,
Food is found where people need it.
All we need is found in granite
Once we have the men to plan it.
Yeast and algae give us meat,
Soil is almost obsolete.
Man can grow to pastures greener
Till all the earth is Pasadena.
Moral:
Man's a nuisance, man's a crackpot,
But only man can hit the jackpot.

There is something to be said for the criticism that conservation involves a negative attitude, not only because the conservationist often appears to put a higher value on non-human materials and their preservation than on man whom they serve, but also because he tends to underestimate the development of man's ability to create resources. It is common sense to conserve renewable resources like soil, water, or the biotic world in general by wise use which allows or encourages replacement of what is used, in cases where the return equals the costs of conservation. It is also intelligent and may soon be essential to use non-renewable resources as carefully and as efficiently as possible, although enough may not be known in any particular situation to decide whether it is best to use or to conserve. But the essence of resources is use, not meaningless restriction. Sound and positive conservation means only that misuse is avoided, and that the environmental base is preserved or permitted to renew itself.

FUTURE RESOURCES

No one can be certain how far or how rapidly technological change or other factors can alter resource use, although the record of the past hundred years in this respect suggests that the rate and degree of change have consistently been underestimated. One can be sure only that the meaning and use of resources are highly changeable and that the future cannot accurately be measured in terms of the present. Man may well be on the threshold of a vast new development in resource use which will add new resources to the list and change the meaning of many present ones. But because a material or advantage is technically usable does not necessarily mean that it is economically so, if total costs of use are too high. It is also generally unwise to assume that when new resources come into being the present locational pattern of economic activity will be greatly distorted. It might be in certain cases, as it clearly was by the advent of coal as the single most important mineral resource of the modern world. But because the Congo has much uranium, for example, does not mean that it will become a great manufacturing or population center on the basis of nuclear power. As with resources in the past and present, assembly and distribution costs will continue to operate. The broad, varied bases of settlement, manufacturing, transport, markets, and capital will continue to be more important than specific mineral deposits in themselves, even or especially if all rocks or all soils should become usable sources of minerals. General technological change is more likely to benefit established centers than presently underdeveloped areas. The economic gap between the developed and underdeveloped areas is unfortunately still widening rather than narrowing, and this trend cannot be reversed merely by the discovery or use of new minerals or power sources in the underdeveloped areas. This is regrettably one of the few things of which one can be relatively sure. The industrial or "developed" areas of the West are still increasing their per capita production and consumption faster than most of the rest of the world, which will thus continue to be "underdeveloped" even if it is able to increase its per capita production. One important factor is the rapid increases in the population of most of the underdeveloped areas as their economic growth proceeds; population increases threaten to consume most or all of the new increments in production so that per capita consumption may rise very little if at all. Another decade or two might see some change in this relationship; Europe's experience with population growth may not be applicable to Asia, for example, but it will be difficult for the presently underdeveloped areas to reach the point where their per capita production increases as rapidly or more rapidly than Europe's at present. Nuclear power may be economically usable on a large scale in competition with coal and oil in the near future, but the advantages for the use of nuclear power are still likely to accrue primarily to the existing industrial areas where all the other necessary factors are concentrated, especially the major markets with their high constant demand for power which helps to keep power costs low. An even more promising development in the long run is the direct use of energy from the sun as heat or in some other form, as plants use it. This still seems to be a long way from being possible economically on a large scale, but if and when it can be achieved, the Sahara Desert will not become the industrial capital of the world.

Additions to the list of usable resources, including many of a less dramatic kind than nuclear or solar energy, may continue to increase the standard of living, and one may hope also the common ground between nations. In a world of relative plenty where goods are better distributed and resources more widely used, there may be less occasion for discord. Such possibilities are of transcendent importance for all of mankind. But resources are not exclusively physical phenomena so much as they are reflections and creations of the human factor. This is apparent in the tremendous regional diversity of the world and in the particular patterns of resource use which have grown up in its different parts. The materials or advantages on which civilizations and economies depend vary enormously. Resource appraisals vary as places, people and societies vary. One of the greatest challenges to modern civilization is whether it can increase and spread resource use widely enough so that, by varying local combinations of resources and wants, economic abundance is available to all.

Questions for Further Study and Discussion

1. On what resources did the pre-Columbian Indians of North America depend? How did regional differences in resource use affect regional differences in culture and vice versa?

2. What current changes in resource use and what new resources can you discern in the contemporary world? in contemporary United States?

3. Why is there so little fishing in Africa? in central and northern South America? Why should the effective market for fish be limited in mainland Asia? Is it likely to increase? Why is fishing so little developed in Australia? Why has South Africa developed so much more of a fishing industry than the rest of Africa? Why is fishing so prominent in the Caribbean, south India and Ceylon, Indonesia, and Madagascar, and yet apparently so unproductive according to the graph of fish catch by countries?

4. Consult an atlas and correlate fisheries with the global pattern of ocean currents and depths.

5. What other specific examples are there of currently unused mineral deposits? Why in each case are they not more fully used? See among other things maps of world iron ore and coal.

6. What does *intervening opportunity* mean? Give other examples of it, and discuss its relevance to matters such as urban growth, retailing, and regional development as well as to resource use.

7. What other major raw materials lose much weight in the manufacturing process? How does this affect the location of the manufacturing in question? the transport of the raw material?

8. What is the national population density (people per square mile) of Norway, Argentina, Holland, Java, Tibet, Yugoslavia, Australia, Japan, Canada, Afghanistan, Malta, and Puerto Rico? Which of these countries are overpopulated? What does overpopulation mean?

9. What are the causes of famine?

10. Why are there only five or six major world centers of heavy manufacturing? Or would you argue that there are more? Why has each of them developed?

11. Compare the location of manufacturing with world maps of coal, iron ore, and petroleum. What do you deduce from the comparison? What do you deduce from comparing the distribution of coal deposits and population?

12. What articles or goods in daily use in the United States have no connection with steel? with refining? How does the supply and cheapness of fresh fruit or ready-made clothing depend on steel?

13. Why is there little or no major manufacturing in the heavily mineralized areas of the Canadian Shield? in the mineralized areas of south-central Australia?

14. In what way does the distribution of predominant economies indicate resource use? To what extent is this general pattern the result of environmental factors?

15. Why do iron and steel manufacturing usually go together? What is pig iron?

16. Why can iron and steel plants on the east coast of the United States afford to import iron ore from Venezuela or Canada when domestic ore deposits are closer? What various factors have to be considered? How may the usability of taconite affect the locational pattern of iron and steel manufacturing? Where are the major east coast plants, and why? How old are they, and why?

17. Find examples of raw material or power sites which once supported manufacturing but where the manufacturing has now moved closer to the market. Why has this happened in each case?

18. If all rocks or soil should become economically usable sources of all minerals, how would this affect the locational pattern of manufacturing?

19. How and in what ways and areas did the use of coal distort the locational pattern of economic activity from the eighteenth century onward? Is such a distortion likely again? Why or why not and under what circumstances?

20. Resource surveys and inventories have been made in great detail for the United States. Is it known precisely what the resources of the United States are or what their quantitative total is? Is the answer subject to change?

21. What is meant by a "currently classified ore"? How and why might such classifications change? Define *ore.*

22. What does the record of American agriculture in the past fifty years suggest about the possibility of increasing total food production and yields for the world as a whole? See Chauncy D. Harris, "Agricultural Production in the United States: The Past Fifty Years and the Next," *Geographical Review,* XLVII (1957), 175-93.

23. What is "wise resource use"? What is "misuse"?

Selected Samples for Further Reading

Alexandersson, G. *Geography of Manufacturing.* Englewood Cliffs, N. J., 1967. A basically informational account of the distribution and evolution of manufacturing complexes.

Apter, E. D. "Nationalism, Government, and Economic Growth," *Economic Development and Cultural Change,* VII (1959), 117–36. A general but very useful discussion of this critically important factor in economic development.

Boesch, H. *A Geography of World Economy.* Princeton, 1964. A stimulating text in economic geography which emphasizes world patterns of resource use.

Brookfield, H. C. *The Geography of Population.* London, 1961. A general text covering the geographic aspect of demography.

Burton, I. and L., and Kates, R. W. *Readings in Resource Management and Conservation.* Chicago, 1965. A collection of articles dealing with the nature of resources, the demographic factor, and the choices confronting the conservationist.

Clarke, J. I. *Population Geography.* London, 1966. A general and topical treatment emphasizing spatial patterns of population.

Coale, A. J., and Hoover, E. M. *Population Growth and Economic Development in Low-Income Countries.* Princeton, 1958. A careful study of the relation between increases in population and increases in production in the areas of the world where this relationship is most crucial.

Dasmann, R. F. *Environmental Conservation.* New York, 1968. An excellent text, stressing the unity of the environment, man's destructive role thus far, and the problems of the future.

Ehrlich, P. R., and Ehrlich, A. H. *Population, Resources, and Environment: Issues in Human Ecology.* San Francisco, 1970. The relation between population, food supply, and the environmental base, stressing the crisis posed by current population increases.

Firey, W. *Man, Mind, and Land: A Theory of Resource Use.* Glencoe, Ill., 1960. A clear discussion of how resources come into being through the interplay of human and physical factors.

Firth, R., and Yamey, B. S., eds. *Capital, Saving, and Credit in Peasant Societies.* Chicago, 1964. Sample studies of the combined force of economic and cultural factors in the operation of communities typical of the underdeveloped world, and the prospects for change.

Fisher, J. L., and Potter, N. *World Prospects for Natural Resources.* Baltimore, 1964. A brief, clear inventory and discussion of projected resource demands and supplies.

Freedman, R., ed. *Population: The Vital Revolution.* Chicago, 1965. An outstanding general treatment by a variety of distinguished demographers.

Galbraith, J. K. *Economic Development.* Cambridge, Mass., 1964. Brief and simple discussion by an eminent economist which summarizes the major problems in economic growth.

Gershenkron, A. *Economic Backwardness in Historical Perspective.*

Cambridge, Mass., 1962. Sample studies of "retarded" development, including pre-Soviet Russia, and some imaginative and valuable ideas.

Ginsburg, N. "Natural Resources and Economic Development," *Annals of the Association of American Geographers*, XLVII (1957), 197–212. A good summary analysis of the role played by non-human resources in economic growth.

———, ed. *Atlas of Economic Development*. Chicago, 1961. Ingenious maps of various "indices" of development, and useful text.

———, ed. *Essays on Geography and Economic Development*. Chicago, 1960. Papers by several geographers.

Hauser, P. M. "Demographic Indicators of Economic Development," *Economic Development and Cultural Change*, VII (1959), 98–116. An excellent analysis, with useful statistics, of the measurement of per capita economic growth.

Herfindahl, O. C. *Natural Resource Information for Economic Development*. Baltimore, 1969. The need for careful resource inventory and evaluation and its role in planning for economic growth.

Highsmith, R. M., Jensen, J. G., and Rudd, R. D. *Conservation in the United States*. 2nd ed. Chicago, 1969. A clearly organized introductory text.

Isaac, E. *Geography of Domestication*. Englewood Cliffs, N. J., 1970. The origins and spread of early food production and animal rearing.

Krause, J. T. "Some Implications of Recent Research in Demographic History," *Comparative Studies in Society and History*, I (1959), 164–88. Discussion of some important additions to our understanding of demographic changes in early modern Europe, with direct bearing on Malthusian theory and on comparisons with other pre-industrial or early-industrial societies.

Landsberg, H. H. *Natural Resources for United States Growth*. Baltimore, 1964. A compact summary in paperback of an earlier and larger work, *Resources in America's Future* (1963), projecting demands and supplies of United States resources to A.D. 2000.

Lord, R. *The Care of the Earth: A History of Husbandry*. New York, 1963. Summarizes especially the growth of agricultural land use in terms of the balance of nature and the concerns of the conservationist.

Mellor, J. *The Economics of Agricultural Development*. Ithaca, 1966. Contemporary world agricultural problems as an aspect of economic growth.

National Academy of Sciences. *Resources and Man*. San Francisco, 1969. A multi-author committee report in eight chapters, presenting a careful analysis of the precarious balance between a booming world population and consumption and world potential for production of food, minerals, and energy.

Park, C. F. *Affluence in Jeopardy: Minerals and the Political Economy*. San Francisco, 1968. A rather polemical but disturbing and well-supported analysis of imminent mineral shortages and the possible consequences in an unbalanced and overpopulated world.

Singer, H. W. *International Development: Growth and Change*. New

York, 1964. Essays by an economist with long and varied field experience, emphasizing in particular the importance of social institutions such as education as preconditions for economic growth, the need for critical resource inventories, and the economic costs of high birth and death rates.

Udall, S. *The Quiet Crisis.* New York, 1963. The growth of the conservation movement in the United States, and some contemporary problems.

Udvardy, M. D. F. *Dynamic Zoogeography, with Special Reference to Land Animals.* New York, 1969. A comprehensive textbook which emphasizes the ecological and historical or developmental aspects of the earth's fauna.

Ward, R. J., ed. *The Challenge of Development: Theory and Practice.* Chicago, 1967. A collection of essays on economic models for growth, and a variety of experiences in applying them.

Williams, T. I., and Derry, T. K. *A Short History of Technology From the Earliest Times to A. D. 1900.* New York, 1961. Based on the authors' five-volume work but not merely an abridgement; an authoritative and sprightly treatment.

Zelinsky, W., ed. *Geography and a Crowding World.* Oxford, 1970. A symposium dealing with population pressures in the developing countries.

Zelinsky, W. *A Prelude to Population Geography.* Englewood Cliffs, N. J., 1966. An outline of the nature and aims of the geographic study of population.

———, Kosinski, L., and Prothero, R. *Essays on Population Pressures upon Resources.* New York, 1970. 32 essays by specialists, focused on the social and economic problems of population growth in the developing countries.

Zimmerman, E. W. *World Resources and Industries.* Rev. Ed. New York, 1951. An excellent detailed discussion of the meaning of resources and a comprehensive survey of world resources and their use. The first ten chapters have been updated and republished as *Introduction to World Resources,* ed. H. H. Hunker, New York, 1965.

Index

Printed in U.S.A.